Memoirs of the American Mathematical Society

Number 196

William Pardon

Local surgery and the exact sequence of a localization for Wall groups

Published by the

AMERICAN MATHEMATICAL SOCIETY

Providence, Rhode Island

VOLUME 12 · ISSUE 2 · NUMBER 196 (end of volume) · NOVEMBER 1977

AMS (MOS) Subject Classification numbers (1970): Primary: 57D65; Secondary: 10C05, 18F25.

Key Words and Phrases: Surgery, Wall group, Hermitian form, $\mathbb{Z}_n$-manifold, stratified space, localization.

Library of Congress Cataloging in Publication Data CIP

Pardon, William, 1947-
Local surgery and the exact sequence of a localization for Wall groups.

(Memoirs of the American Mathematical Society ; no. 196)
"Volume 12, issue 2."
Bibliography: p.
Includes index.
1. Surgery (Topology) 2. Group rings. 3. Localization theory. I. Title. II. Series: American Mathematical Society. Memoirs ; no. 196.
QA3.A57 no. 196 [QA613.658] 510'.8s [514'.7]
ISBN 0-8218-2196-2 77-11963

Table of Contents

Introduction v

Abstract xii

Chapter One : Local surgery and the exact sequence of a localization 1

Chapter Two : The category $\mathfrak{D}_F^1$ of torsion modules with short free resolution and its L-groups 10

Chapter Three : Moore spaces 29

Chapter Four : A class of stratified spaces and some of its geometric properties 37

Part A: Geometry of Moore spaces 38

Part B: Conglomerate Moore spaces 55

Part C: Composition of conglomerate Moore spaces 80

Chapter Five : The $(\pi - \pi)$-theorem 100

Chapter Six : Local surgery in the odd-dimensional case 117

Chapter Seven : Local surgery in the even-dimensional case 127

Chapter Eight : The simply-connected case and π_1 unrestricted 159

Index of Symbols and Terminology 167

Bibliography 169

To my parents

Introduction

The rational group ring $\mathbb{Q}\pi$ of a finite group π is relatively well-understood through the classical Maschke theorem ($\mathbb{Q}\pi$ is a product of simple algebras) and the Wedderburn theorem (each simple algebra is a full matrix ring over a division ring). As a result, constructions functorial in π and depending only on $\mathbb{Q}\pi$ are equally well-understood. For example, $K_0(\mathbb{Q}\pi)$ is the rational representation ring of π (studied in classical character theory), $K_1(\mathbb{Q}\pi)$ is the direct product of certain "positive" subgroups of abelianized unit groups of the division rings appearing in the above-mentioned decomposition of $\mathbb{Q}\pi$, and $L_*(\mathbb{Q}\pi)$ (the Wall groups) are discussed in [W3] (see also [P4]). On the other hand, such complete results are not available in case we replace $\mathbb{Q}\pi$ by $\mathbb{Z}\pi$, the integral group ring. In contrast to our precise knowledge of the ring structure of $\mathbb{Q}\pi$, much less is known about that of $\mathbb{Z}\pi$. But interest in the groups $K_i(\mathbb{Z}\pi)$ and $L_i(\mathbb{Z}\pi)$ (for any finitely-presented group π) is correspondingly greater because of their connection with finiteness up to homotopy ($K_0(\mathbb{Z}\pi)$), simple homotopy ($K_1(\mathbb{Z}\pi)$), pseudo-isotopy ($K_2(\mathbb{Z}\pi)$) and diverse questions in manifold theory ($L_*(\mathbb{Z}\pi)$).

The way to formalize the relation between $K_i(\mathbb{Z}\pi)$ and $K_i(\mathbb{Q}\pi)$ is through "the exact sequence of a localization", due, for low values of i to Heller and Reiner [HR] and, in general, to Quillen [Gr]. To state this result one needs to introduce a category $\mathfrak{D}^1$ whose objects are finitely-generated, $\mathbb{Z}$-torsion, $\mathbb{Z}\pi$-modules M having homological dimension one; i.e., having a short projective resolution

(I.1) $$P \rightarrowtail Q \twoheadrightarrow M$$

(I.2) <u>Theorem</u> (Quillen). There is an exact sequence of abelian groups for all $i \geq 0$

$$K_{i+1}(\mathbb{Z}\pi) \longrightarrow K_{i+1}(\mathbb{Q}\pi) \longrightarrow K_i(\mathfrak{D}^1) \longrightarrow K_i(\mathbb{Z}\pi) \longrightarrow K_i(\mathbb{Q}\pi).$$

Here $K_i(\mathfrak{D}^1)$ is constructed in [Gr] and $K_0(\mathfrak{D}^1)$ is the Grothendieck group based on the free abelian monoid with basis the isomorphism classes $\{M\}$ of objects M of $\mathfrak{D}^1$, modulo the relation $\{M\} = \{M'\} + \{M''\}$ where $M' \rightarrowtail M \twoheadrightarrow M''$ is a short exact sequence.

We think of $K_i(\mathfrak{D}^1)$ as the local term in (I.2) because each object M of $\mathfrak{D}^1$ is canonically the direct sum $\Sigma_p M_{(p)}$ of its localization $M_{(p)}$ for all primes p in $\mathbb{Z}$.

The first theorem of this paper ((1.11)) is a geometric construction of the localization sequence for the Wall groups $L^h_*(\mathbb{Z}\pi)$, $L^h_*(\mathbb{Q}\pi)$ (hereafter the superscript "h" is understood for all Wall groups).

(I.3) Theorem. For each $i \geq 0$ and any CW complex K there is an abelian group $L_i^{\mathbb{Q}/\mathbb{Z}}(K)$, functorial in K, and an exact sequence

$$L_{i+1}(\mathbb{Z}\pi) \longrightarrow L_{i+1}(\mathbb{Q}\pi) \longrightarrow L_i^{\mathbb{Q}/\mathbb{Z}}(K) \longrightarrow L_i(\mathbb{Z}\pi) \longrightarrow L_i(\mathbb{Q}\pi).$$

Let us explain how $L_n^{\mathbb{Q}/\mathbb{Z}}(K)$ is defined and the theorem is proved. Recall from [W2, §9] that we may regard $L_n(\mathbb{Z}\pi)$ as a set of cobordism classes of surgery problems (degree-one normal maps) $(f,b)\colon (N,\nu_N) \to (Y,\xi)$ together with a reference map $\omega\colon Y \to K$, where N is an n-manifold, Y a Poincaré complex, ξ any bundle over Y, $f|\partial N\colon \partial N \to \partial Y$ is a homotopy equivalence and $\pi = \pi_1 K$. Similarly, requiring that $f|\partial N$ only induce a rational homology equivalence (with π-local coefficients) and allowing $f|\partial N$ to vary through "rational h-cobordisms" (in the cobordism relation) we obtain $L_n(\mathbb{Q}\pi)$ (see (1.7)). It is now almost immediate that, in order to prove (I.3), we should define $L_i^{\mathbb{Q}/\mathbb{Z}}(K)$ to be rational h-cobordism classes of normal maps $(f,b)\colon (M^i,\nu_M) \to (X,\xi)$, where f induces a rational homology equivalence and $f|\partial M$ is a homotopy equivalence.

Even though (I.3) is a trivial consequence of correctly formulated definitions, there remains the interesting question of identifying the groups $L_i^{\mathbb{Q}/\mathbb{Z}}(K)$ as the K-groups in some category (analogous to (I.2)) where we think of $L_i(\mathbb{Z}\pi)$ (resp. $L_i(\mathbb{Q}\pi)$) as the K-groups in a category whose objects are nonsingular Hermitian (quadratic) forms on free $\mathbb{Z}\pi$-modules (resp. free $\mathbb{Q}\pi$-modules). This can be done because the groups $L_i^{\mathbb{Q}/\mathbb{Z}}(K)$ turn out to be surgery obstruction groups.

To begin the program of identification, we call a surgery problem $(f,b)\colon (N,\nu_N) \to (Y,\xi)$ <u>local</u> if f induces a rational homology equivalence, or (in other words) $K_*(N)$ is $\mathbb{Z}$-torsion, where $K_*(N)$ is the kernel of the map on homology induced by f; <u>local</u> <u>bordisms</u> are defined similarly. (We have justified the adjective "local" above; (I.3) is also related to the "local-to-global" Hasse-Minkowski principle--see the introduction to [P3]). To construct a local "surgery theory" we first observe that each local surgery problem is locally bordant to a highly-connected one (1.19). The proof of this suggests that we have to take bordisms with handles in two adjacent dimensions as the "basic" local surgery. Indeed, it is easily seen that killing a homology class x of finite order in dimension $k - 1$, creates a class y of infinite order in dimension k, so that we no longer have a local surgery problem; but killing any multiple of y yields a local surgery problem and the union of the bordisms corresponding to the above surgeries is a local bordism. Examining this procedure geometrically motivates the main geometric construction in this paper, the conglomerate Moore space (a stratified space with two strata). Namely, if we arrange that the image S^k of the attaching map for the second surgery meets the handle for the first surgery in core k-discs, $D_1^k,\ldots,D_t^k$ (see (4B.54) and (4B.57))

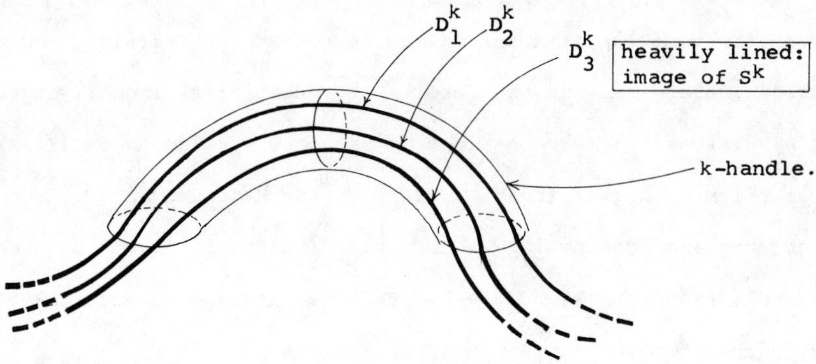

we see that the pair of surgeries is determined by $S^k - (\mathring{D}_1^k \cup \cdots \cup \mathring{D}_t^k)$. To make this fact useful homologically and to keep track of framings, we replace $S^k - (\mathring{D}_1^k \cup \cdots \cup \mathring{D}_t^k)$ by the Moore space $\mathfrak{M}_t^k$ obtained by identifying its t boundary components to a single $(k-1)$-sphere (denoted $b\mathfrak{M}_t^k$ and called the bockstein) by orientation-preserving homeomorphisms; observe that $b\mathfrak{M}_t^k$ carries the homology class x. $\mathfrak{M}_t^k$ is a $\mathbb{Z}_t$-manifold (see [BRS]) and $\mathfrak{M}_t^k - b\mathfrak{M}_t^k$ and $b\mathfrak{M}_t^k$ are manifolds, so we think of $\mathfrak{M}_t^k$ as a stratified space. ($\mathfrak{M}_t^k$ is not a Moore space in the usual sense since $\pi_1\mathfrak{M}_t^k \neq 0$ (see (3.2))). In summary, a "basic" local surgery is a surgery (in some sense) on a framed imbedding of the space $\mathfrak{M}_t^k$, instead of on a framed imbedded sphere. The "conglomerate" Moore space in (4B.1) is needed to carry out a completely general discussion; this space is essentially a connected sum of copies of the $\mathfrak{M}_t^k$, for various t.

We can now describe the first four chapters of this paper. Chapter One makes precise the above informal proof of (I.3), Chapter Two records the algebraic definitions and properties of the groups $L_n(\mathbb{Q}\pi/\mathbb{Z}\pi)$ (the local surgery obstruction groups) and Chapter Three develops some "mod t" homotopy theory. Chapter Four, by far the longest, supplies constructions for (conglomerate) Moore spaces, analogous to those used for spheres in ordinary surgery theory: normal bundle and immersion ((4A.15), (4B.18)) "immersion theory" ((4A.21)), fibre over a point in a trivial bundle ((4A.19), (4B.31)) and intersection theory ((4A.29), (4A.39),

(4B.33), (4B.45)). A more complete summary of the contents of these chapters can be found at the beginning of each and a list of symbols and terminology is at the end of the paper. Beyond Chapter One it is always assumed that fundamental groups of all manifolds are finite. (However, in Chapter Eight we indicate how this restriction can be removed.)

Next this machinery is used to develop the local surgery theory: Chapter Five furnishes the $(\pi - \pi)$-theorem, and Chapters Six and Seven identify the local surgery obstruction groups in the odd-and even-dimensional cases, respectively. For example, to each highly-connected local surgery problem $(f,b): (M^{2k+1}, \nu_M) \to (Y,\xi)$ is associated its linking form on the k-dimensional kernel; and f is locally bordant to a homotopy equivalence if and only if the form is hyperbolic in the appropriate sense. This kernel lies in $\mathfrak{D}^1_F$, the category whose objects are $\mathbb{Z}$-torsion, $\mathbb{Z}\pi$-modules with short free resolution (compare (I.1)), in analogy to the K-theoretic localization of (I.2): the surgery obstruction group $L_0(\mathbb{Q}\pi/\mathbb{Z}\pi)$ is, roughly speaking, K_0 of a suitable category of hermitian (quadratic) forms on objects in $\mathfrak{D}^1_F$. In a similar way, the obstruction group in dimension 2k can be thought of as K_1 of the above category of forms over $\mathfrak{D}^1_F$. Just as in [W2, §9] the information developed in Chapters Five, Six, and Seven suffices to identify the groups $L^{\mathbb{Q}/\mathbb{Z}}_i(K)$ in (I.3) with the corresponding local surgery obstruction groups $L_*(\mathbb{Q}[\pi_1 K]/\mathbb{Z}[\pi_1 K])$, $* = 0,1$. (Only L_0 and L_1 are used to denote the local surgery obstruction groups: periodicity shows we do not need to deal with "higher" K-functors (K_i, $i \geq 2$) and we wish to preserve the analogy with (I.2)). It is remarkable that the geometric analysis in these chapters can be carried out largely by following [W2, §4,5,6] except that conglomerate Moore spaces must be substituted for spheres. Finally, Chapter Eight computes the local surgery obstruction groups in the simply-connected case and shows how to remove the finiteness restriction on π_1.

Let us indicate here how the above ideas have been applied

elsewhere. First, the localization sequence (I.3) is difficult to apply to computations of $L_n(\mathbb{Z}\pi)$ (π finite), largely because a list of indecomposable objects in the category $\mathfrak{T}^1_F$ is unknown (even given knowledge of $K_0(\mathbb{Z}\pi)$). However, many finite groups (including finite cyclic groups, generalized quaternion groups, and metacyclic groups) can be broken down by means of Cartesian squares into components, each of which is a ring of integers in a number field (or a non-commutative ring having similar ideal-theoretic properties); and if in (I.3), $\mathbb{Z}\pi$ is replaced by such a ring, and $\mathbb{Q}\pi$ by its fraction field (or ring), the local term in (I.3) becomes easily computable. (Such rings appear geometrically in the work of Petrie [Pe].) There are Mayer-Vietoris sequences for the above Cartesian squares ([Ba] and [P2]) while the required algebraic generalization of the localization sequence turns out to be a consequence of the geometric work in this paper: it is sufficient to observe how the geometric data in (I.3) are translated to algebra by Chapters Six and Seven. The details are carried out in [P3]; [P5] applies Dress induction, Sharpe's periodicity theorem [Sh 1] and this more general localization sequence to calculations of Wall groups. Max Karoubi [K], Andrew Ranicki [R], and Gunnar Carlsson and R. James Milgram [CM] have also produced algebraic versions of the localization sequence, independently of the author and of one another. Jean-Claude Hausmann [Ha] has used Dress induction and the localization sequence to show that $L_{2k+1}(\mathbb{Z}\pi) \to L_{2k+1}(\mathbb{Q}\pi)$ (see (I.3)) is the zero map for all finite groups π and all k.

On the geometric side, Justin Smith has studied more general localizations in [Sm]. He obtains results toward the classification of Poincaré embeddings, showing in particular that the knot concordance group is a local surgery obstruction group. Neal Stoltzfus [St] has treated knots (imbeddings $S^n \to S^{n+2}$) and knot concordance groups in a formal way, analogous to that in which we have treated surgery problems and $L_n(\mathbb{Z}\pi)$. He thus obtains a "rational knot concordance group," a "torsion (local) knot concordance group," and a localization sequence relating the three groups. Ted Petrie [Pe] (also Browder and Quinn) has developed a general theory of equivariant normal maps where the group action is not necessarily free. The orbit space of such an action is stratified by orbit types, and "stratum-by-stratum" surgery leads to a sequence of obstructions in the groups $L_*(\mathbb{Z}_{(p)}\pi)$. When $\mathbb{Z}_{(p)}\pi$ has good homological properties (e.g., when $(p,|\pi|) = 1$), these obstructions can be computed using (I.3). Finally, we mention that the idea of "surgery on a conglomerate Moore space" (4B.53) can be pushed much further. Conglomerates are constructed (in (4B.1)) "according to a resolution" of length two (like (I.1) where P and Q are $\mathbb{Z}\pi$-free). This can be generalized to the construction of stratified spaces corresponding to any chain complex and leads to an a priori geometric surgery obstruction, which realizes the algebraic formulation of Ranicki [Ra 2]. In [BRS] bordism theories with coefficients "in a resolution" are constructed in almost the same way. However, in their context the resolution is fixed while in ours it is not. An early version of a part of the localization sequence (although it was not recognized as such) was studied in [C]. We have borrowed some ideas from this work.

I would like to thank A.J.C. for constant encouragement while this paper was being revised. Ms. Kate March did the typing as fast and accurately as one could want.

This paper is a revised version of the author's doctoral dissertation, written at Princeton University under the direction of William Browder whom I wish to thank for his guidance. I am grateful to the National Science Foundation for its support both while I was a graduate student, and later while this paper was being revised.

William Pardon

Abstract

Let $L_n^h(A)$ denote the Wall group in dimension n, of the ring-with-involution A. Then the exact sequence of a localization relates $L_n^h(\mathbb{Z}\pi)$ to $L_n^h(\mathbb{Q}\pi)$ in an exact sequence of the form

$$\cdots \longrightarrow L_{n+1}^h(\mathbb{Z}\pi) \longrightarrow L_{n+1}^h(\mathbb{Q}\pi) \longrightarrow L_n^h(\mathbb{Q}\pi/\mathbb{Z}\pi) \longrightarrow L_n^h(\mathbb{Z}\pi) \longrightarrow L_n^h(\mathbb{Q}\pi) \longrightarrow \cdots$$

where π is a finite group and $L_n^h(\mathbb{Q}\pi/\mathbb{Z}\pi)$, the local term, is the obstruction group for local surgery. Geometrically, $L_n^h(\mathbb{Q}\pi/\mathbb{Z}\pi)$ is the group of bordism classes of normal maps $f: N^n \to Y^n$ with $f_*: H_*(N;\mathbb{Q}\pi) \overset{\cong}{\to} H_*(Y;\mathbb{Q}\pi)$, where normal bordisms satisfy a similar condition. Exact sequences of this kind are well-known in algebraic K-theory, where the "local" and "global" terms ($L_n^h(\mathbb{Q}\pi/\mathbb{Z}\pi)$ and $L_n^h(\mathbb{Q}\pi)$ here) are used to gain information about the "integral" term, $L_n^h(\mathbb{Z}\pi)$.

By developing the properties of a space $\mathfrak{M}_t^k$, called a Moore space, and of a more complicated stratified space $\mathcal{C}_\mu^k$, a conglomerate Moore space, it turns out that one may study the surgery theory $L_*^h(\mathbb{Q}\pi/\mathbb{Z}\pi)$ in such a way that the spaces $\mathcal{C}_\mu^k$ play the same role that spheres play in "ordinary" surgery theory $L_*^h(\mathbb{Z}\pi)$. This allows one to give an algebraic description of $L_n^h(\mathbb{Q}\pi/\mathbb{Z}\pi)$ analogous to that of $L_n^h(\mathbb{Z}\pi)$: the definition of $L_n^h(\mathbb{Z}\pi)$ is based on hermitian forms over free $\mathbb{Z}\pi$-modules, while that of $L_n^h(\mathbb{Q}\pi/\mathbb{Z}\pi)$ is based on forms over $\mathbb{Z}$-torsion $\mathbb{Z}\pi$-modules which have a short $\mathbb{Z}\pi$-free resolution.

Received by the editors March 5, 1975 and in revised form January 13, 1977.

Chapter One: Local Surgery and the Exact Sequence of a Localization

Chapter One begins with the definitions of cobordism groups ${}^{i}L_n^{\mathbb{Q}/\mathbb{Z}}(K)$ of "local normal maps", i = 1,2, where K is a CW complex, connected if i = 2. ((1.5) and (1.6)) Discussion of these groups is analogous to that of the groups $L_n^i(K)$ in [W2,§9] so detailed definitions are given, but proofs are omitted. The main result is the localization sequence (1.11), which follows formally from our definitions (see Remark following (1.11)). The unknown term ${}^{1}L_n^{\mathbb{Q}/\mathbb{Z}}(K)$, in the exact sequence will be evaluated in later chapters (in case $\pi_1 K$ is finite) by studying local surgery problems (1.12), and this chapter concludes with their simplest property: each local surgery problem may be assumed "highly-connected", up to local bordism ((1.19),(1.20)). In this chapter no restriction is placed on $\pi_1 K$.

1.1. Let $\mathbb{Z}\pi$ denote the integral group ring of the group π and $\mathbb{Q}\pi$, its rational group ring. Following [W2] all $\mathbb{Z}\pi$-modules M will be right $\mathbb{Z}\pi$-modules; $M \otimes \mathbb{Q}\pi$ denotes the right $\mathbb{Q}\pi$-module $M \otimes_{\mathbb{Z}\pi} \mathbb{Q}\pi$. If X is a CW complex, $H_*(X)$ denotes integral (cellular) homology; if X is equipped with a right cellular π-action, then $H_*(X)$ is a right $\mathbb{Z}\pi$-module.

1.2 Definition. $\theta = \{(g;b): (N,\partial N;\nu_N) \to (Y,X;\xi),\ \omega: Y \to K\}$ is an n-dimensional local object if

(a) $(g;b): (N,\partial N;\nu_N) \to (Y,X;\xi)$ is a normal map where $g|\partial N$ is a homotopy equivalence and dim N = n,

(b) K is a CW complex and $\omega: Y \to K$ is a continuous map, and

(c) $\hat{g}_* \otimes \mathbb{Q}\pi: H_*(\hat{N}) \otimes \mathbb{Q}\pi \to H_*(\hat{Y}) \otimes \mathbb{Q}\pi$ is an isomorphism of $\mathbb{Q}\pi$-modules, where $\pi = \pi_1 K$ and $\hat{g}: \hat{N} \to \hat{Y}$ is the map covering g, induced by ω from the universal covering $\widetilde{K} \to K$.

1.3 Definition. Let θ be an n-dimensional local object. Then θ is null-bordant provided

(a) there is a normal map $(G;B): (P,N,N';\nu_P) \to (Z,Y,Y';\Xi)$ where $\dim P = n+1$, $\partial P = N \cup N'$, $N \cap N' = \partial N$, $(G,B)|(N,\nu_N) = (g,b)$ and $G|N': N' \to Y'$ is a homotopy equivalence,

(b) there is a map $\Omega: Z \to K$ extending ω, and

(c) $\hat{G}_* \otimes \mathbb{Q}\pi: H_*(\hat{P}) \otimes \mathbb{Q}\pi \to H_*(\hat{Z}) \otimes \mathbb{Q}\pi$ is an isomorphism of $\mathbb{Q}\pi$-modules where $\pi = \pi_1 K$ and $\hat{G}: \hat{P} \to \hat{Z}$ is the map covering G, induced by Ω from the universal covering $\widetilde{K} \to K$.

1.4 Definition. If θ is a local object, $(-\theta)$ denotes θ with the orientations of N and Y reversed; $\underline{0}$ denotes the object where $N = \emptyset = Y$. Let $\theta_1 \sim \theta_2$ if $\theta_1 + (-\theta_2)$ is null-bordant, where "+" means disjoint union.

1.5 Proposition. "$\sim$" is an equivalence relation on the set of n-dimensional local objects. The equivalence classes form an abelian group ${}^1L_n^{\mathbb{Q}/\mathbb{Z}}(K)$, functorial in K, where addition is induced by disjoint union and the zero element is represented by $\underline{0}$.

Proof. See [W2, 9.1].

1.6 Definition-Proposition. Let K be a connected complex and let ${}^2L_n^{\mathbb{Q}/\mathbb{Z}}(K)$ be the set of restricted local objects modulo the equivalence relation "$\approx$" where

(a) a restricted local object is a local object (1.2) where $\omega_\#: \pi_i Y \xrightarrow{\cong} \pi_i K$, $i = 0,1$, and

(b) if θ_1 and θ_2 are restricted local objects, $\theta_1 \approx \theta_2$ if $\theta_1 + (-\theta_2)$ is null-bordant where $\Omega_\#: \pi_i Z \xrightarrow{\cong} \pi_i K$, $i = 0,1$ (in (1.3)(b)).

Define $\theta_1 \approx 0$ by removing the symbol θ_2 from the condition in (b) above. (There is no zero object here.) Objects θ such that $\theta \approx 0$ constitute an equivalence class and ${}^2L_n^{\mathbb{Q}/\mathbb{Z}}(K)$ is a pointed set, functorial in K.

Proof. [W2, 9.2].

1.7 Definition. $\theta = \{(g;b): (N,\partial N;\nu_N) \to (Y,X;\xi),\ \omega: Y \to K\}$ is an n-dimensional rational object if in (1.2)(a) the condition that $g|\partial N$ be a homotopy equivalence is replaced by the condition that $(\hat{g}|\partial\hat{N})_* \otimes \mathbb{Q}\pi: H_*(\partial\hat{N}) \otimes \mathbb{Q}\pi \to H_*(\hat{X}) \otimes \mathbb{Q}\pi$ be an isomorphism (where $\hat{g}$ is defined in (1.2)(c)), (1.2)(b) is retained, and (1.2)(c) is omitted. θ is null-bordant provided

(a) there is a normal map $(G;B): (P,N,N',R;\nu_P) \to (Z,Y,Y',W;\Xi)$ and a continuous map $\Omega: Z \to K$ extending ω such that $\dim(P) = n + 1$, $\partial P = N \cup R \cup N'$, $R \cap (N \cup N') = \partial R = \partial N \cup \partial N'$, $(G,B)|(N,\nu_N) = (g,b)$, and $(\hat{G}|\hat{N}')_* \otimes \mathbb{Q}\pi: H_*(\hat{N}') \otimes \mathbb{Q}\pi \to H_*(\hat{Y}') \otimes \mathbb{Q}\pi$ is an isomorphism where $\hat{G}: \hat{P} \to \hat{Z}$ is the map covering G induced by $\Omega: Z \to K$ from the universal covering $\tilde{K} \to K$, and

(b) $(\hat{G}|\hat{R})_* \otimes \mathbb{Q}\pi: H_*(\hat{R}) \otimes \mathbb{Q}\pi \cong H_*(\hat{W}) \otimes \mathbb{Q}\pi$.

Let θ_1 and θ_2 be n-dimensional rational objects. Let $\theta_1 \sim \theta_2$ if $\theta_1 + (-\theta_2)$ is null-bordant. The next proposition follows by the argument of [W2, 9.1].

1.8 Proposition. "$\sim$" is an equivalence relation whose equivalence classes ${}^1L_n^{\mathbb{Q}}(K)$ form an abelian group functorial in K.

1.9. We leave to the reader the formulation of the definition of the pointed set ${}^2L_n^{\mathbb{Q}}(K)$. Recall from [W2, §9] that $L_n^1(K)$ is defined by omitting condition (c) from (1.2) and (1.3); $L_n^2(K)$ is defined analogously. We denote these groups ${}^1L_n^{\mathbb{Z}}(K)$ and ${}^2L_n^{\mathbb{Z}}(K)$, respectively. It is important to know they are the same since the latter is related to surgery (compare (1.22)) while the former has good formal properties (see (1.11)).

Following [W2,9.4], we have an analogous result.

1.10 Proposition. If K is a connected CW complex and $n \geq 5$, the forgetful map induces natural isomorphisms of pointed sets,

$$ {}^2L_n^{\mathbb{Q}/\mathbb{Z}}(K) \xrightarrow{\cong} {}^1L_n^{\mathbb{Q}/\mathbb{Z}}(K) \quad \text{and} \quad {}^2L_n^{\mathbb{Q}}(K) \xrightarrow{\cong} {}^1L_n^{\mathbb{Q}}(K) . $$

The next theorem is the first form of the main result in this paper (see the Remark following its proof).

1.11 Theorem (The Exact Sequence of a Localization). Let K be a CW complex. There is an exact sequence of abelian groups, functorial in K, where $n > 1$,

$$ \cdots \longrightarrow {}^1L_{n+1}^{\mathbb{Q}}(K) \xrightarrow{d} {}^1L_n^{\mathbb{Q}/\mathbb{Z}}(K) \xrightarrow{i} {}^1L_n^{\mathbb{Z}}(K) \xrightarrow{r} {}^1L_n^{\mathbb{Q}}(K) \xrightarrow{d} {}^1L_{n-1}^{\mathbb{Q}/\mathbb{Z}}(K) \longrightarrow \cdots $$

Proof. The map i is defined by forgetting condition (1.2)(c); r is defined by relaxing the boundary condition (see (1.7)); and d is defined by taking the boundary of a rational object. It is clear from the definitions of the groups involved that these maps are well-defined. Exactness and naturality also follow easily from the definitions.

Remark. The formulation and proof of the localization sequence are completely formal. Its content is the algebraic determination of the groups and maps involved, in terms of classical invariants. Indeed, one of the main results of [W2] is the isomorphism ${}^1L_n^{\mathbb{Z}}(K) \overset{\cong}{\to} L_n(\mathbb{Z}[\pi_1 K])$ ($n \geq 5$), while it follows easily from [CS] that ${}^1L_n^{\mathbb{Q}}(K) \cong \Gamma_n(\mathbb{Z}\pi \to \mathbb{Q}\pi)$ $= L_n(\mathbb{Q}[\pi_1 K])$ ($n \geq 5$). Our main result is the introduction of a "local surgery theory" furnishing (in (6.1) and (7.1)) analagous isomorphisms ${}^1L_n^{\mathbb{Q}/\mathbb{Z}}(K) \overset{\cong}{\to} L_n(\mathbb{Q}[\pi_1 K]/\mathbb{Z}[\pi_1 K])$ in case $\pi_1 K$ is finite and $n \geq 9$. ($L_n(\mathbb{Q}\pi/\mathbb{Z}\pi)$ is defined in Chapter Two.)

There are notions of local surgery and local bordism corresponding to the groups ${}^iL_n^{\mathbb{Q}/\mathbb{Z}}(K)$ introduced above. First recall from [W2] that if

$(h;b)\colon (M,\partial M;\nu_M) \to (Y,X;\xi)$ is a normal map, then $K_i(M)$ (resp. $K^i(M)$) denotes the kernel (resp. cokernel) of $\hat{h}_*\colon H_i(\hat{M}) \to H_i(\tilde{Y})$ (resp. $\hat{h}^*\colon H^i(\tilde{Y}) \to H^i(\hat{M})$) where $\hat{h}\colon \hat{M} \to \tilde{Y}$ is the map covering h, induced by the universal covering $\tilde{Y} \to Y$; $K_i(M,\partial M)$ and $K^i(M,\partial M)$ are defined analogously.

1.12 <u>Definition</u>. (a) A normal map $(g;b)\colon (N,\partial N;\nu_N) \to (Y,X;\xi)$ is a <u>local surgery problem</u> if $K_*(N) \otimes \mathbb{Q}\pi \equiv 0$, where $\pi = \pi_1 Y$. (b) Let $(G;B)\colon (P,N,N',R;\nu_P) \to (Y \times I, Y \times \{0\}, Y \times \{1\}, X \times I;\Xi)$ be a normal bordism between $(G;B)|(N,\nu_N)$ and $(G;B)|(N',\nu_{N'})$. Then $(G;B)$ is a <u>local bordism</u> if $K_*(P) \otimes \mathbb{Q}\pi \equiv 0 \equiv K_*(R) \otimes \mathbb{Q}\pi$; it is a local bordism <u>relative to ∂N</u> if, in addition, $R = (\partial N) \times I$ and $G|R = (G|\partial N) \times I$.

A basic property of normal maps is that, up to normal cobordism, they may be assumed "highly connected". The same is true of local surgery problems (1.19)(b) although the result is slightly different. To carry out the surgery we need some simple properties of $\mathbb{Z}\pi$-modules.

If C is a chain complex, free and finitely generated over $\mathbb{Z}\pi$ in each dimension, there are isomorphisms (of chain complexes),

1.13 (a) $C \otimes_{\mathbb{Z}} \mathbb{Q} \cong C \otimes \mathbb{Q}\pi$, (b) $\mathrm{Hom}_{\mathbb{Z}\pi}(C,\mathbb{Z}\pi) \otimes \mathbb{Q}\pi \xrightarrow{\cong} \mathrm{Hom}_{\mathbb{Z}\pi}(C,\mathbb{Q}\pi)$,

1.14 $$C \otimes \mathbb{Q}\pi \xrightarrow{\cong} \mathrm{Hom}_{\mathbb{Q}\pi}(\mathrm{Hom}_{\mathbb{Z}\pi}(C,\mathbb{Q}\pi),\mathbb{Q}\pi).$$

Setting $H_*(C;\mathbb{Q}\pi) = H_*(C \otimes \mathbb{Q}\pi)$ and $H^*(C;\mathbb{Q}\pi) = H_*(\mathrm{Hom}(C,\mathbb{Q}\pi))$, the universal coefficient theorem and (1.13)(a)-(b) yield isomorphisms

1.15 (a) $H_*(C) \otimes \mathbb{Q}\pi \xrightarrow{\cong} H_*(C;\mathbb{Q}\pi)$, (b) $H^*(C) \otimes \mathbb{Q}\pi \xrightarrow{\cong} H^*(C;\mathbb{Q}\pi)$.

Moreover, (1.13)(a) holds for an arbitrary $\mathbb{Z}\pi$-module M in place of C (use the isomorphism $M \xrightarrow{\cong} M \otimes \mathbb{Z}\pi$) showing that

1.16 tensoring with $\mathbb{Q}\pi$ preserves exactness.

1.17 <u>Lemma.</u> Let $S = \mathbb{Z}\pi$ or $\mathbb{Q}\pi$. Let D be a chain complex of finite length, free and finitely generated over S in each dimension. Suppose that for some n, $H_i(D) = 0$, $i < n$. Then there is a natural isomorphism $H^n(D) \xrightarrow{\cong} \mathrm{Hom}_S(H_n(D),S)$.

Proof. Follow the proof of [CS,1.4] using right-exactness of $\mathrm{Hom}_S(-,S)$.

1.18 Lemma. Let $(g;b)\colon (N,\partial N;\nu_N) \to (Y,X;\xi)$ be a local surgery problem. Then $K_*(N,\partial N) \otimes \mathbb{Q}\pi \equiv 0 \equiv K_*(\partial N) \otimes \mathbb{Q}\pi$.

Proof. Let C denote the chain complex such that $H_*(C) = K_*(N,\partial N)$, and let $D = \mathrm{Hom}_{\mathbb{Z}\pi}(C,\mathbb{Q}\pi)$. By assumption and Poincaré duality [W2,§2], $K^*(N,\partial N) \otimes \mathbb{Q}\pi \equiv 0$. Thus, by (1.15)(b), $H_*(D) \equiv 0$, and by (1.17) $H^*(D) \equiv 0$. By (1.14) and (1.15)(a), $H_*(C) \otimes \mathbb{Q}\pi \equiv 0$. The exact sequence of the pair $(N,\partial N)$ shows $K_*(\partial N) \otimes \mathbb{Q}\pi \equiv 0$.

1.19 Proposition. Let $(g;b)\colon (N,\partial N;\nu_N) \to (Y,X;\xi)$ be a local surgery problem where $\dim(N) \geq 6$, and $g|\partial N$ is a homotopy equivalence.

(a) If $(G;B)\colon (P,N,N',R;\nu_P) \to (Y\times I, Y\times\{0\}, Y\times\{1\}, X\times I;\xi\times I)$ is a local bordism of $(g;b)$, then $(G;B)|(N',\partial N';\nu_{N'})$ is a local surgery problem.

(b) $(g;b)$ is locally bordant (relative to ∂N) to $(g';b')\colon (N',\partial N';\nu_{N'}) \to (Y,X;\xi)$ where $g'_\#\colon \pi_1 N' \xrightarrow{\cong} \pi_1 Y$, and

$$K_i(N') = \begin{cases} 0 & \text{if } i \neq k, k-1 \text{ and } \dim(N) = 2k \\ 0 & \text{if } i \neq k \qquad \text{and } \dim(N) = 2k+1. \end{cases}$$

Proof. We omit bundle data from the notation.

(a): $(G;B)$ and $(G;B)|R$ are by definition local surgery problems, so by (1.18), $K_*(\partial P) \otimes \mathbb{Q}\pi \equiv 0 \equiv K_*(\partial R) \otimes \mathbb{Q}\pi$. Since $\partial P = R \cup (N \cup N')$ where $R \cap (N \cup N') = \partial R$, a Mayer-Vietoris sequence shows $K_*(N') \otimes \mathbb{Q}\pi \equiv 0$.

(b): We may assume Y is connected. Let $\pi = \pi_1 Y$. Since $K_0(N) \otimes \mathbb{Q}\pi = 0$ while $K_0(N)$ is torsion-free as an abelian group, $K_0(N) = 0$. There is a short exact sequence $\pi_1\hat{N} \rightarrowtail \pi_1 N \overset{g_\#}{\twoheadrightarrow} \pi_1 Y$ ($g_\#$ is surjective because

g is of degree one), where $\pi_1\hat{N}$ is normally generated in $\pi_1 N$ by a finite number of elements [Br, IV.1.15]. Let $G'\colon (W',N',N) \to (Y\times I, Y\times\{1\}, Y\times\{0\})$ be the bordism corresponding to surgery on these elements. We have the exact sequence

$$K_2(N) \rightarrowtail K_2(W') \xrightarrow{j} K_2(W',N) \twoheadrightarrow K_1(N),$$

and an isomorphism $(G'|N')\colon \pi_1 N' \overset{\cong}{\to} \pi_1 Y$. Since $K_*(N) \otimes \mathbb{Q}\pi = 0$, (1.16) shows $j \otimes \mathbb{Q}\pi$ is an isomorphism. Thus $K_2(W') \otimes \mathbb{Q}\pi$ is $\mathbb{Q}\pi$-free and since $K_2(W')$ is $\mathbb{Z}\pi$-finitely generated, there is a finitely generated free $\mathbb{Z}\pi$-module E and an injection

$$i\colon E \rightarrowtail K_2(W')$$

such that $i \otimes \mathbb{Q}\pi\colon E \otimes \mathbb{Q}\pi \overset{\cong}{\to} K_2(W') \otimes \mathbb{Q}\pi$. Since $\dim(N) \geq 6$, the inclusion $N' \to W'$ induces an isomorphism $\ell\colon K_2(N') \overset{\cong}{\to} K_2(W')$. Surgery on framed imbeddings $S^2 \to N'$ representing the image under $\ell^{-1}i$ of a basis of E yields a bordism of $G'|N'$, $G''\colon (W'',N'',N') \to (Y\times I, Y\times\{1\}, Y\times\{0\})$ such that

(a) $K_i(N'') = 0$, $i = 0,1$,

(b) $G = (G' \cup G'')\colon (W' \cup W'', N'', N) \to (Y\times I, Y\times\{1\}, Y\times\{0\})$ is a local bordism of g to $G|N''$, relative to ∂N,

(c) $G|N''$ is a local surgery problem.

Part (a) is clear by construction and (c) follows from (b) and (1.18). To prove (b) consider the exact sequence

$$K_3(W' \cup W'',N) \rightarrowtail K_3(W'',N') \xrightarrow{\partial} K_2(W',N) \twoheadrightarrow K_2(W' \cup W'',N)$$

where by construction ∂ is induced by (more precisely, is isomorphic to) $ji\colon E \to K_2(W',N)$. Since $(ji) \otimes \mathbb{Q}\pi$ is an isomorphism, $K_i(W' \cup W'',N) \otimes \mathbb{Q}\pi = 0$, $i = 2,3$. This shows $K_*(W' \cup W'',N) \otimes \mathbb{Q}\pi \equiv 0$ and since $K_*(N) \otimes \mathbb{Q}\pi \equiv 0$, $K_*(W' \cup W'') \otimes \mathbb{Q}\pi \equiv 0$.

Continuing in this way we may arrange that $(g;b)$ be locally bordant (relative to $g|\partial N$) to $(g';b'): (N',\partial N';\nu_{N'}) \to (Y,X;\xi)$ where

$$K_i(N') = \begin{cases} 0, & \text{if } i < k-1 \text{ and } \dim N' = 2k \\ 0, & \text{if } i < k \text{ and } \dim N' = 2k+1. \end{cases}$$

In case $\dim N' = 2k$, Poincaré Duality and the fact that $K^i(N') = 0$, $i < k - 1$, finish the proof. If $\dim N' = 2k + 1$, $K_i(N') = 0$, $i > k + 1$, by the same argument; $K_{k+1}(N') \cong K^k(N') \cong \mathrm{Hom}_{\mathbb{Z}\pi}(K_k(N'),\mathbb{Z}\pi)$ where the latter isomorphism is by (1.18). Since $K_k(N') \otimes \mathbb{Q}\pi \cong K_k(N') \otimes_{\mathbb{Z}} \mathbb{Q}$ (1.15) and $\mathbb{Z}\pi$ is $\mathbb{Z}$-torsion free, $\mathrm{Hom}_{\mathbb{Z}\pi}(K_k(N'),\mathbb{Z}\pi) = 0$. This completes the proof.

The proof of the following relative version of (1.19) is left to the reader.

1.20 Proposition. Let $(g;b): (N,\partial N;\nu_N) \to (Y,X;\xi)$ be a local surgery problem where $\dim N \geq 6$. Then $(g;b)$ is locally bordant to $(g';b'): (N',\partial N';\nu_{N'}) \to (Y,X;\xi)$ where the non-zero kernels of $(g';b')$ appear in the exact sequence

(a) $K_k(N') \rightarrowtail K_k(N',\partial N') \to K_{k-1}(\partial N') \to K_{k-1}(N') \twoheadrightarrow K_{k-1}(N',\partial N')$ when $\dim N = 2k$,

(b) $K_k(\partial N') \rightarrowtail K_k(N') \to K_k(N',\partial N') \twoheadrightarrow K_{k-1}(\partial N')$ when $\dim N = 2k + 1$,

and $g'_{\#}: \pi_1 N' \overset{\cong}{\to} \pi_1 Y$, $(g'|\partial N')_{\#}: \pi_1(\partial N') \overset{\cong}{\to} \pi_1 X$.

1.21 Definition. The local surgery problem $(g';b')$ in (1.19)(b) is said to be highly connected.

Finally, we can relate ${}^2L_n^{\mathbb{Q}/\mathbb{Z}}(K)$ to local surgery by observing that a restricted local object becomes a local surgery problem after we forget $\omega: Y \to K$ in (1.6)(a).

1.22 Theorem. Let $\theta \in {}^2L_n^{\mathbb{Q}/\mathbb{Z}}(K)$, where $\pi_1 K$ is finite and $n \geq 9$. Then $\theta \approx 0$ (1.6) if and only if a local surgery problem in the class of θ is locally bordant to a homotopy equivalence.

Proof. This is a consequence of the "(π-π) Theorem" (5.3), just as [W2,9.3] is a consequence of [W2,3.3].

Chapter Two: The Category $\mathfrak{D}_F^1$ of torsion modules with short free resolution and its L-groups.

This chapter contains the definitions and main examples necessary to study the obstruction groups for torsion-surgery (1.12). It is divided into three parts: the first gives just enough information to define (2.13) the group $L_0^\epsilon(B/A)$ of ϵ-hermitian forms on torsion modules; the second ((2.15)-(2.31)) discusses this definition to prepare for the geometry of Chapter Six (and may be omitted on first reading); the third gives the definition (2.39) of $L_1^\epsilon(B/A)$, the group of ϵ-hermitian "formations" ([Ra 1],[Nov]) on torsion modules. The algebra of this chapter is treated more fully in [P3,§2]. It is assumed that the reader is familiar with the basic language of quadratic forms and surgery theory. References are [Ba] and [W2], respectively.

2.1. In this chapter A will always denote the integral group ring $\mathbb{Z}\pi$ of a finite group π. A is a ring-with-involution, denoted "___", defined by $\overline{\sum_{g\in\pi} n_g g} = \sum_{g\in\pi} n_g g^{-1}$ ($n_g \in \mathbb{Z}$) and satisfying $\overline{a+b} = \bar{a} + \bar{b}$, $\overline{ab} = \bar{b}\bar{a}$, and $\bar{1} = 1$, for all $a,b \in A$. Let B denote $\mathbb{Q}\pi$, the rational group ring; B/A is an (non-finitely-generated) A-module.

2.2 Definitions. By an A-module M we always mean a right A-module, unless otherwise specified. The torsion part τM of M is the A-submodule of $\mathbb{Z}$-torsion elements of M; denote $M/\tau M$ by fM. If M is an A-module, then $\mathrm{Hom}_A(M,A)$ is a left A-module in a natural way; it is converted to an A-module, denoted $\bar{M}$, by setting $(ga)(m) = \bar{a}g(m)$, $g \in \mathrm{Hom}_A(M,A)$, $a \in A$, $m \in M$. The left A-module $\mathrm{Hom}_A(M,B/A)$ is similarly converted to an A-module, denoted $M^\wedge$. Given an A-module homomorphism $g\colon M \to N$, there are homomorphisms $\bar{g}: \bar{N} \to \bar{M}$ and $g^\wedge: N^\wedge \to M^\wedge$ defined in the usual way. A homomorphism $\alpha: E \to F$ between free A-modules with chosen bases will be identified with its matrix, also denoted α; conversely, any rectangular matrix with entries in A yields a homomorphism of free A-modules. The induced homomorphism $\bar{\alpha}: \bar{F} \to \bar{E}$ has (in the dual basis) matrix equal to the conjugate transpose of μ, also denoted $\bar{\mu}$.

2.3 Convention. From now on all A-modules except B/A and B/J_ϵ (defined in (2.5d)) will be $\mathbb{Z}$-finitely-generated and, hence, A-finitely-generated.

2.4 Proposition. (a) Given an A-module M, there is an A-module structure on $Hom_{\mathbb{Z}}(M,\mathbb{Z})$ (resp. $Hom_{\mathbb{Z}}(M,\mathbb{Q}/\mathbb{Z})$) and a natural A-isomorphism $\bar{G}: \bar{M} \xrightarrow{\cong} Hom_{\mathbb{Z}}(M,\mathbb{Z})$ (resp. $G^\wedge: M^\wedge \xrightarrow{\cong} Hom_{\mathbb{Z}}(M,\mathbb{Q}/\mathbb{Z})$). (b) The map $M \to fM$ (2.2) induces an isomorphism $\overline{(fM)} \xrightarrow{\cong} \bar{M}$. (c) If $M' \to M \to M''$ is a short exact sequence of A-modules, the induced sequence $(M'')^\wedge \to M^\wedge \to (M')^\wedge$ is short exact; if, in addition, $\tau M'' = 0$, $(\bar{M}'') \to \bar{M} \to (\bar{M}')$ is short exact. (d) If $\tau M = 0$ (resp. $fM = 0$) there is a natural isomorphism $M \xrightarrow{\cong} \bar{\bar{M}}$ (resp. $M \xrightarrow{\cong} M^{\wedge\wedge}$).

Proof. (a) $Hom_{\mathbb{Z}}(M,\mathbb{Z})$ is an A-module by setting $(ga)(m) = g(m\bar{a})$, $g \in Hom_{\mathbb{Z}}(M,\mathbb{Z})$, $a \in A$, $m \in M$. Define $\bar{G}: \bar{M} \to Hom_{\mathbb{Z}}(M,\mathbb{Z})$ by $\bar{G}(h) = h(m)_e$ = coefficient (in $\mathbb{Z}$) of e (the identity in π) on $h(m) \in A = \mathbb{Z}\pi$. It is easy to show $\bar{G}$ is a natural isomorphism; $G^\wedge$ is treated similarly. Thus statements (b)-(d) are translated by (a) into statements about $\mathbb{Z}$-modules, which are trivial.

We will now study forms.

2.5 Definition. Let M, N be A-modules and let V = A, B, or B/A.

(a) A function $\varphi: M \times N \to V$ is A-sesquilinear if, for all $m,m' \in M$, $n,n' \in N$, $a,b \in A$,

$$\varphi(ma + m', nb + n') = \bar{a}\varphi(m,n)b + \bar{a}\varphi(m,n') + \varphi(m',n)b + \varphi(m',n').$$

(b) Such a φ is <u>nonsingular</u> if the adjoint of φ, $Ad(\varphi): M \to \bar{N}$ (or $N^\wedge$) is an isomorphism $((Ad(\varphi))(m)(n) = \varphi(m,n))$. (c) If in (a) $M = N$, φ is <u>ϵ-hermitian</u>, $\epsilon = \pm 1$, if $\varphi(m,n) = \overline{\epsilon\varphi(n,m)}$, $n, m \in M$. (d) Let $J_\epsilon = \{a + \epsilon\bar{a} \mid a \in A\}$ (an involution invariant $\mathbb{Z}$-submodule of A), let M be an A-module and let $\psi: M \to B/J_\epsilon$ be a function. The triple (M,φ,ψ) is called an <u>ϵ-linking form</u> if for $m,n \in M$ and $a \in A$,

(i) $\varphi: M \times M \to B/A$ is ϵ-hermitian and sesquilinear

(ii) $\psi(m) \equiv \varphi(m,m)$ (mod A)

(iii) $\psi(m+n) \equiv \psi(m) + \psi(n) + \varphi(m,n) + \overline{\epsilon\varphi(m,n)}$ (mod J_ϵ)

(iv) $\psi(ma) = \bar{a}\psi(m)a$.

2.6 <u>Examples</u>. If X is a CW complex, we follow the convention of [W2, Chap. 2] by writing $H_*(X)$ for the $\mathbb{Z}[\pi_1 X]$-module $H_*(\tilde{X})$, $\tilde{X}$ the universal cover of X. If (P^n,g,F) is a normal map [W2, p. 9], so that $g: P^n \to X$ is a degree-one map of the oriented n-manifold P^n to the Poincare complex X, then $K_i(g)$ or $K_i(P)$ denotes the kernel of $g_*: H_i(P) \to H_i(X)$. We have the following examples [W1, W2]($\pi_1 P = \pi_1 X$ finite.)

(a) In (2.5b) let $M = fK_i(P)$, $N = fK_{n-i}(P)$, $V = A$ and φ the intersection pairing [W1]; again in (2.5b) let $M = \tau K_i(P)$, $N = \tau K_{n-i-1}(P)$, $V = B/A$, and φ the linking pairing. Poincare duality implies the pairings are non-singular.

(b) Let $\dim P = n = 2k + 1$ with $K_i(P) = 0$, $i < k$. Then if $M = \tau K_k(P)$, [W1, Thm. 5.2] gives an example of a (nonsingular) ϵ-linking form, $\epsilon = (-1)^{k+1}$.

(c) Let S be a torsion A-module ($\tau S = S$). There is a sesquilinear form $\nu: S^\wedge \times S \to B/A$, the <u>natural form</u>, defined by $\nu(f,s) = f(s)$, where $f \in S^\wedge$ and $s \in S$. $Ad(\nu): S^\wedge \to S^\wedge$ is the identity map by its definition. Hence ν is nonsingular.

2.7 <u>Definition</u>. Let $\mathfrak{D}_F^1$ denote the category of torsion A-modules having short free resolution over A; hence $S \in \mathfrak{D}^1_F$ if $S \otimes_A B = 0$ and there is a short exact sequence

$$E \rightarrowtail F \twoheadrightarrow S$$

where E and F are A-free (it is clearly enough to require E and F to be stably free). As a topological example we have:

2.8 Proposition. Let C_* be a chain complex of free A-modules with $C_i = 0$, $i < 0$ and $i > N$, and $H_i(C_*) = 0$, $i \neq k$, $k \geq 2$. Then $H_k(C_*) \in \mathfrak{D}^1_F$.

Proof. Truncating C_* below dimension k we obtain the exact sequence $C_N \rightarrowtail \ldots \to C_{k+1} \to Z_k \twoheadrightarrow H_k(C_*)$, where Z_k is the A-module of k-cycles. Z_k is stably free over A by induction (see [W2, p. 26]), and so a theorem of Rim [R, Thm. 4.12] completes the proof.

2.9 Definition. An ϵ-form is an ϵ-linking form (2.5 (d)) (S,φ,ψ) where φ is nonsingular and $S \in \mathfrak{D}^1_F$ (2.7). Two ϵ-forms are isometric if there is a φ- and ψ-preserving isomorphism between them. (Note that an ϵ-form (S,φ,ψ) always has φ nonsingular, whereas an ϵ-linking form need not have this property.)

2.10 Example. If in (2.6b) we assume $fK_k(P) = 0$, Poincare duality implies $K_i(P) = 0$, $i > k$ (use (2.6a)). Apply Prop. (2.8) to C_* with $C_i = C_{i+1}(M_{\tilde{g}},\tilde{P})$, $\tilde{P}$ the universal cover of P, $\tilde{g}\colon \tilde{P} \to \tilde{X}$ the corresponding map, and $M_{\tilde{g}}$ the mapping cylinder of $\tilde{g}$. The ϵ-linking form of (2.6b) is thus an ϵ-form since $K_k(P) = H_{k+1}(M_g,P)$, $\varepsilon = (-1)^{k+1}$.

2.11 Definition. Let (S,φ,ψ) be an ϵ-form. (a) (S,φ,ψ) is a kernel, if there is a submodule $K \subset S$, $K \in \mathfrak{D}^1_F$ such that $\varphi \mid K \times K \equiv 0 \equiv \psi \mid K$ and $K^\perp = K$, where $K^\perp = \{s \in S \mid \varphi(s,K) = 0\}$. K is called a subkernel. (b) (S,φ,ψ) is hyperbolic if $S = U^\wedge + U$, $U \in \mathfrak{D}^1_F$, $\varphi \mid U \times U \equiv 0 \equiv \psi \mid U$, $\varphi \mid U^\wedge \times U^\wedge \equiv 0 \equiv \psi \mid U^\wedge$, and $\varphi \mid U^\wedge \times U$

is the natural form (2.6(c)). (S,φ,ψ) is determined up to isometry by the isomorphism class of U and is denoted $\mathscr{H}(U)$.

2.12 Remarks. (a) If (S,φ,ψ) is a kernel with subkernel K, there is an induced nonsingular form $\varphi': S/K \times K \to B/A$ obtained by restricting φ to K in the second variable and using $\varphi(K,K) \equiv 0$ to divide out by K in the first. To see φ' is nonsingular, observe first that $\mathrm{Ad}(\varphi')$ is injective because $K = K^{\perp}$. Any element of $K^{\wedge}$ comes by restriction from an element of $S^{\wedge}$ (use the isomorphism $K^{\wedge} \cong \mathrm{Hom}_{\mathbb{Z}}(K,\mathbb{Q}/\mathbb{Z})$ in (2.4a)), so the surjectivity of $\mathrm{Ad}(\varphi)$ and $K = K^{\perp}$ imply $\mathrm{Ad}(\varphi')$ is surjective.
(b) If $|T|$ denotes the number of elements in the torsion A-module T, then in the setting of (a), $|K|^2 = |S|$. This follows from the nonsingularity of φ' and the fact that $|K| = |K^{\wedge}|$ (again using (2.4a)). (c) If $A = \mathbb{Z}$, $B = \mathbb{Q}$, $\epsilon = 1$, $S = \mathbb{Z}/p^2\mathbb{Z} = \langle s\rangle$, p an odd prime, $\varphi(s,s) = 2p^{-2} \in \mathbb{Q}/\mathbb{Z}$ and $\psi(s) = 2p^{-2} \in \mathbb{Q}/2\mathbb{Z}$, then (S,φ,ψ) is a kernel, but clearly cannot be hyperbolic. This exemplifies the difference between kernels and hyperbolic ϵ-forms.

2.13 Theorem-Definition. Let $F_0^{\epsilon}(B/A)$ denote the set of isometry classes of ϵ-forms. Orthogonal sum defines on $F_0^{\epsilon}(B/A)$ the structure of an abelian semigroup; let $\mathfrak{F}_0^{\epsilon}(B/A)$ denote the corresponding Grothendieck group, and $\mathscr{K}^{\epsilon}$ the subgroup generated by kernels. Then $\mathfrak{F}_0^{\epsilon}(B/A)/\mathscr{K}^{\epsilon}$ is an abelian group, denoted $L_0^{\epsilon}(B/A)$. The class of (S,φ,ψ) in $L_0^{\epsilon}(B/A)$ is denoted $[S,\varphi,\psi]$.

Proof. Given (S_1,φ_1,ψ_1), $(S_2,\varphi_2,\psi_2) \in F_0^{\epsilon}(B/A)$ define their (orthogonal) sum by $(S_1,\varphi_1,\psi_1) \perp (S_2,\varphi_2,\psi_2) = (S_1 + S_2,\varphi_1 + \varphi_2,\psi_1 + \psi_2)$ where on the right side we mean the usual "orthogonal sum" of forms. If the zero of $F_0^{\epsilon}(B/A)$ is represented by (S,φ,ψ) with $S = (0)$, $F_0^{\epsilon}(B/A)$ is an abelian semigroup. Now $\mathfrak{F}_0^{\epsilon}(B/A)$ is (by definition) the free abelian group on isometry classes, modulo the relation

$$\{(S_1,\varphi_1,\psi_1) \perp (S_2,\varphi_2,\psi_2)\} - (S_1,\varphi_1,\psi_1) - (S_2,\varphi_2,\psi_2) = 0.$$

The final statement is obvious.

2.14 <u>Proposition</u>. Let (S,b,q) represent $\alpha \in L_0^\epsilon(B/A)$. Let β be represented by $(S,-b,-q)$ where $(-b)(s,t) = -b(s,t), (-q)(s) = -q(s)$, $s,t \in S$. Then $\alpha + \beta = 0$ in $L_0^\epsilon(B/A)$.

<u>Proof</u>. Let $K \subset S \times S$ be the image of the diagonal inclusion $S \to S + S$. Then K is annihilated by $b + (-b)$ and $q + (-q)$ and $K = K^\perp$. Hence $\alpha + \beta$ is represented by $(S + S,\ b + (-b),\ q + (-q))$, a kernel.

Readers familiar with the functor $\tilde{K}_0$ of an abelian category $\mathcal{C}$, $\tilde{K}_0\mathcal{C}$, will recognize the connection between it and $L_0^\epsilon(B/A)$, for appropriate $\mathcal{C}$. We will not make this precise (it could be done); but we observe that it is now natural to define $L_1^\epsilon(B/A)$, analogous to $K_1\mathcal{C}$. In order to make our definition of $L_1^\epsilon(B/A)$ seem more natural and to simplify the geometry of Chapter Six, we will weaken the requirement for an ϵ-form (V,ζ,ξ) to represent zero in $L_0^\epsilon(B/A)$. Before stating the result (2.22) which does this, we need some propositions which will also be used in later chapters.

Given a torsion-free (resp. torsion) A-module M, the notion of "dual module" $\bar{M}$ (resp. $M^\wedge$) has been defined (see (2.2), (2.4d)). The following proposition relates the torsion-free to the torsion case.

2.15 <u>Proposition</u>. If $M \in \mathfrak{D}_F^1$ (2.7) has short free resolution $E \overset{\mu}{\rightarrowtail} F \overset{j}{\twoheadrightarrow} M$, there is a "dual" short free resolution of $M^\wedge$,

$$\bar{F} \xrightarrow{\bar{\mu}} \bar{E} \xrightarrow{\tilde{j}} M^\wedge .$$

<u>Proof</u>. This is in [W1, p.249], but we need a formula for $\tilde{j}$, so a proof will be sketched. Applying $\mathrm{Hom}_A(-,A)$ to the short free resolution of M yields the exact sequence

$$\bar{F} \xrightarrow{\bar{\mu}} \bar{E} \xrightarrow{j'} \mathrm{Ext}_A^1(M,A)$$

Identifying $\mathrm{Ext}^1_A(M,A)$ with $\mathrm{cok}(\bar{\mu})$, define an isomorphism k: $\mathrm{cok}(\bar{\mu}) \to M^\wedge$ as follows. Let $f \in \bar{E}$. The bottom horizontal map can be defined to make the following diagram commute because B is A-injective

2.16
$$\begin{array}{ccc} E & \xrightarrow{\quad f \quad} & A \\ \mu \downarrow & & \downarrow \text{inclusion} \\ F & \xrightarrow{\; f(\mu^{-1}) \;} & B \end{array}$$

The map $f(\mu^{-1})$ is unique because $\mu \otimes B$ is an isomorphism. If $m \in M$ and $x \in \mathrm{cok}(\bar{\mu})$, define k by the formula $k(x)(m) = r \cdot f(\mu^{-1})(j^{-1}(m))$ where $r\colon B \to B/A$ is the quotient map, $j'(f) = x$ and $j^{-1}(m) \in F$ is any element for which $j(j^{-1}(m)) = m$. Thus,

2.17
$$\tilde{j}(f)(m) = rf(\mu^{-1})(j^{-1}(m)).$$

Details are left to the reader.

Forms on torsion A-modules are difficult to study because the classification of torsion A-modules (even those in $\mathcal{T}^1_F$) is largely unknown and in any case quite difficult. (To appreciate this, see [Sz] for a classification when $A = \mathbb{Z}_p$, p prime.) Hence we lift to a torsion-free context (2.19) introduced in [Co]. First a lemma.

2.18 <u>Lemma</u> [Co,1.4]. Let (S,φ,ψ) be an ϵ-form. For each $t \in S$ there is $b \in B$ such that $r'b = \psi(t)$ and $b = \epsilon\bar{b}$, where $r'\colon B \to B/J_\epsilon$ is the quotient map.

<u>Proof</u>. By (2.5b(iv)), $\psi(-t) = \psi(t)$ and by (2.5b(iii)), $\psi(t+(-t)) - \psi(t) - \psi(-t) = -(\varphi(t,t) + \epsilon\overline{\varphi(t,t)})$ so $2\psi(t) = \varphi(t,t) + \epsilon\overline{\varphi(t,t)}$. Hence if $b,b' \in B$ are such that $r'b = \psi(t)$ and $rb' = \varphi(t,t)$ $(r\colon B \to B/A)$, then $2b = b' + \epsilon\bar{b}' + c$, where $c \in J_\epsilon$. Hence $b = 1/2(b' + \epsilon\bar{b}) + 1/2\, c$, so $b = \epsilon\bar{b}$.

With (S,φ,ψ) an ϵ-linking form and $(R) = (E \overset{\mu}{\rightarrowtail} F \overset{j}{\twoheadrightarrow} S)$ a resolution of

S, let $\{f_1,\ldots,f_n\}$ be a basis for F and choose (by (2.18)) $\tau_{ik} \in B$, $1 \leq i,k \leq n$, such that $\varphi(jf_i,jf_k) = r\tau_{ik}$, $\psi(jf_i) = r'\tau_{ii}$ and $\tau_{ik} = \epsilon\bar{\tau}_{ki}$. Use the $(n\times n)$-matrix $\tau = (\tau_{ik})$ to define in the obvious way an ϵ-hermitian form, also denoted τ,

$$\tau\colon F \times F \longrightarrow B.$$

The pair (R,τ) is called a <u>covering</u> of (S,φ,ψ). This terminology is justified by the following proposition, in which part (a) is proved above.

2.19 <u>Proposition</u>. Given an ϵ-linking form (S,φ,ψ) and a resolution $E \overset{\mu}{\rightarrowtail} F \overset{j}{\twoheadrightarrow} S$, there exists an ϵ-hermitian form $\tau\colon F \times F \to B$ such that if $r\colon B \to B/A$ and $r'\colon B \to B/J_\epsilon$ are the projections, $f,f' \in F$ and $e,e' \in E$

(a)
$$\varphi(jf,jf') = r\tau(f,f')$$
and
$$\psi(jf) = r'\tau(f,f).$$

(b) The sesquilinear forms $\tau_\mu\colon F \times E \to B$ and ${}_\mu\tau_\mu\colon E \times E \to B$ defined by $\tau_\mu(f,e) = \tau(f,\mu(e))$ and ${}_\mu\tau_\mu(e,e') = \tau(\mu(e),\mu(e'))$ both take values in A. Also ${}_\mu\tau_\mu$ is ϵ-hermitian and ${}_\mu\tau_\mu(e,e) \in J_\epsilon$.

(c) The diagram

2.20
$$\begin{array}{ccc} F & \xrightarrow{\ \mathrm{Ad}(\tau_\mu)\ } & \bar{E} \\ \Big\downarrow j & & \Big\downarrow \tilde{j} \\ S & \xrightarrow{\ \mathrm{Ad}(\varphi)\ } & S^{\wedge} \end{array}$$

commutes (see (2.15) for $\tilde{j}$), and the matrix of $\mathrm{Ad}(\tau_\mu) = \epsilon\bar{\mu}\tau$ (choosing a basis $\{e_1,\ldots,e_n\}$ for E to give an $(n\times n)$-matrix, also denoted μ, for $\mu\colon E \to F$, and using the dual basis $\{\bar{e}_1,\ldots,\bar{e}_n\}$ for $\bar{E}$).

<u>Proof</u>. (b): $r\,\tau_\mu(f,e) = r\,\tau(f,\mu(e)) =$ (by (a)) $\varphi(jf,j\mu(e))$. But $j\mu \equiv 0$ so $\tau_\mu(f,e) \in \ker(r) = A$. Clearly ${}_\mu\tau_\mu$ takes values in A and is ϵ-hermitian; $r'\,{}_\mu\tau_\mu(e,e) =$ (by (a)) $\psi(j\mu(e))$. Since $j\mu \equiv 0$, ${}_\mu\tau_\mu(e,e) \in \ker(r') = J_\epsilon$. (c): Let $jf = s$. Then

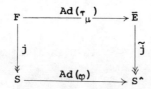

$$\tilde{j}((\mathrm{Ad}\,\tau_\mu)(f'))(s) = (\text{by } (2.17))\ r((\mathrm{Ad}\,\tau_\mu)(f')(\mu^{-1}))(f)$$
$$= r((\mathrm{Ad}\,\tau_\mu)(f'))(\mu^{-1}f) = r\,\tau_\mu(f',\mu^{-1}f)$$
$$= r\,\tau(f',f) = (\text{by (a)})\ \varphi(jf',jf) = \varphi(jf',s)$$
$$= ((\mathrm{Ad}\,\varphi)(jf'))(s).$$

Hence for each $f' \in F$, $\tilde{j}((\mathrm{Ad}\,\tau_\mu)(f')) = \mathrm{Ad}\,\varphi(jf') \in S^\wedge$, which shows that the diagram (2.20) commutes. Computation of the matrix of $\mathrm{Ad}(\tau_\mu)$ is left to the reader; for details see [P3, 1.10], and [Ba, I.2.7].

The following converse to (2.19) is left to the reader.

2.21 Proposition. Let $S \in \mathfrak{D}^1_F$ have short free resolution $E \overset{\mu}{\rightarrowtail} F \overset{j}{\twoheadrightarrow} S$ and let $\tau: F \times F \to B$ be an ϵ-hermitian form satisfying the conditions of (2.19b). Then the equations of (2.19a) define an ϵ-linking form (S,φ,ψ) satisfying (2.19c).

Now we are ready to prove the weak criterion for triviality in $L^\epsilon_0(B/A)$.

2.22 Theorem. An ϵ-form (V,ζ,ξ) represents zero in $L^\epsilon_0(B/A)$ if and only if there is a hyperbolic ϵ-form $\mathscr{H}(U)$ such that $(V,\zeta,\xi) \perp \mathscr{H}(U)$ is a kernel.

Proof. By (2.15) $[V,\zeta,\xi] = 0$ in $L^\epsilon_0(B/A)$ if and only if there are kernels (S,φ,ψ), (S',φ',ψ') and an isometry

2.23 $$(V,\zeta,\xi) \perp (S,\varphi,\psi) \cong (S',\varphi',\psi').$$

Thus the condition in (2.22) is sufficient. Proof of necessity requires the lengthy argument from here to (2.31). Begin by adding $(S,-\varphi,-\psi)$ to both sides of (2.23) to be able to assume (2.23) holds where (S,φ,ψ) has a subkernel $K \subseteq S$ which is an A-direct summand. (The summand K is the "diagonal".) The idea of the proof is to find a summand of S complementary to K, which is also a subkernel.

Let $L \subseteq S$ be such that $L + K = S$. Using $S/K \cong L$ and (2.12a) shows that φ induces a nonsingular form $\varphi'\colon K \times L \to B/A$. Thus

2.24 $$\mathrm{Ad}(\varphi')\colon K \xrightarrow{\cong} L^{\wedge}\ ,\quad \varphi'((\mathrm{Ad}\,\varphi')^{-1}(g),t) = g(t)$$

where $g \in L^{\wedge}$, $t \in L$. Since $K,S \in \mathfrak{D}^1_F$, it follows that $L \in \mathfrak{D}^1_F$, so there is a short free resolution

$$(R') = (E \rightarrowtail^{\alpha} F \xrightarrow{j}\!\!\twoheadrightarrow L)$$

and the corresponding dual resolution $(\bar{R}') = (\bar{F} \overset{\bar{\alpha}}{\rightarrowtail} \bar{E} \overset{\tilde{j}}{\twoheadrightarrow} L^{\wedge})$ for $L^{\wedge}$ (2.15). Choose for this discussion bases $\{e_i\}$ for E, $\{f_i\}$ for F and let $\{\bar{e}_i\},\{\bar{f}_i\}$ be the dual bases for $\bar{E},\bar{F}$; here $i = 1,\ldots,n$. Having made these choices no notational distinction will be made between a homomorphism (or sesquilinear form) involving $E,F,\bar{E}$ or $\bar{F}$, and its corresponding matrix. With obvious notation,

2.25 $$(R) = (E + \bar{F} \xrightarrow{\mu = \begin{pmatrix}\alpha & 0\\ 0 & \bar{\alpha}\end{pmatrix}} F + \bar{E} \xrightarrow{\begin{pmatrix} j & 0\\ 0 & k\end{pmatrix}}\!\!\twoheadrightarrow L + K = S)$$

is a short free resolution for S, where $k = (\mathrm{Ad}\,\varphi')^{-1}\tilde{j}$ (2.24). Let (R',λ) be a covering (2.19) of $(L,\varphi|L,\psi|L)$ and let $\alpha^{-1}\colon \bar{E} \times F \to B$ and $\epsilon\bar{\alpha}^{-1}\colon F \times \bar{E} \to B$ denote the sesquilinear forms with matrices α^{-1} and $\epsilon\bar{\alpha}^{-1}$, respectively. Let $\tau\colon (F + \bar{E}) \times (F + \bar{E}) \to B$ be the ϵ-hermitian form whose matrix is

2.26 $$\tau = \begin{bmatrix} \lambda & \epsilon\bar{\alpha}^{-1} \\ \alpha^{-1} & 0 \end{bmatrix} .$$

2.27 <u>Lemma</u>. (R,τ) is a covering of (S,φ,ψ).

<u>Proof</u>. It is easy to verify that, for R given by (2.25) and τ by (2.26), the condition of (2.21) is satisfied. Hence (R,τ) covers some ϵ-form on S. By construction, the zero in the lower right corner of τ, and λ in the upper left cover the correct forms $(K,\varphi|K,\psi|K)$ and $(L,\varphi|L,\psi|L)$, respectively. It remains to show that if $f_i \in F$ and

$\bar{e}_\ell \in \bar{E}$ are elements in the chosen bases, then $\varphi(k\bar{e}_\ell, j^*f_i) = r(\alpha^{-1})_{\ell i}$, where $(\alpha^{-1})_{\ell i}$ is the (ℓ,i)-entry in the matrix α^{-1} and $r\colon B \to B/A$. Indeed, $\varphi(k\bar{e}_\ell, jf_i) = \varphi'(k\bar{e}_\ell, jf_i)$ = (by definition of k) $\varphi'((\operatorname{Ad}\varphi')^{-1}\widetilde{j\bar{e}_\ell}, jf_i)$ = (by (2.24)) $(\widetilde{j\bar{e}_\ell})(jf_i)$ = (by (2.17)) $r\,\bar{e}_\ell(\alpha^{-1})(f_i) = r(\alpha^{-1})_{\ell i}$. This completes the proof.

By (2.18), $\lambda_{ii} = \lambda_i + \epsilon\bar{\lambda}_i$, for some $\lambda_i \in B$, $i = 1,\ldots,n$. Let λ_s be a splitting of λ:

$$(\lambda_s)_{ij} = \begin{cases} \lambda_{ij} & , \quad i > j \\ 0 & , \quad i < j \\ \lambda_i & , \quad i = j \ . \end{cases}$$

2.28 <u>Lemma</u>. If the $(n\times n)$-matrix $\lambda_s\alpha$ has entries in A, then (S,φ,ψ) is hyperbolic.

<u>Proof</u>: Temporarily changing the given basis of $F + \bar{E}$ by the automorphism $G\colon F + \bar{E} \to F + \bar{E}$ whose matrix (in the given basis) is $\begin{pmatrix} I & 0 \\ \overline{-\lambda_s\alpha} & I\end{pmatrix}$, changes the matrix of τ to

$$\tau' = \begin{pmatrix} I & -\lambda_s\alpha \\ 0 & I \end{pmatrix} \begin{pmatrix} \lambda & \epsilon\bar{\alpha}^{-1} \\ \alpha^{-1} & 0 \end{pmatrix} \begin{pmatrix} I & 0 \\ \overline{-\lambda_s\alpha} & I \end{pmatrix} = \begin{pmatrix} 0 & \epsilon\bar{\alpha}^{-1} \\ \alpha^{-1} & 0 \end{pmatrix} .$$

The automorphism G induces an automorphism g (and thus a change of coordinates) of L + K and it is straightforward to verify that g(L) is a complementary subkernel to K (g(L) is covered by the zero in the upper left corner of τ'). This completes the proof.

Since (R',λ) covers $(L,\varphi|L,\psi|L)$, $\lambda\alpha = \epsilon\overline{\bar{\alpha}\lambda}$ has entries in A (2.19c); in general, this is not true of $\lambda_s\alpha$, so we cannot apply (2.28) directly to conclude (S,φ,ψ) is hyperbolic. But by modifying (R,τ) we will change (S,φ,ψ) so that (2.28) does apply and $(V,\zeta,\xi) \perp (S,\varphi,\psi)$ remains a kernel.

Given $(T,\rho,\sigma) \in F_0^\epsilon(B/A)$ and $J \subseteq T$ such that $J \subseteq J^\perp$, by a construction analogous to that of (2.12a), there are induced nonsingular sesquilinear forms $(T/J^{\perp}) \times J \to B/A$ and $\rho_1: (J^{\perp}/J) \times (J^{\perp}/J) \to B/A$. If, in addition, $\sigma(J) \equiv 0$, there is an induced ϵ-form $(J^{\perp}/J,\rho_1,\sigma_1)$.

2.29 Lemma. Let (T,ρ,σ) be an ϵ-form and let $J \in \mathfrak{D}^1_F$ be a submodule of T such that $J \subseteq J^{\perp}$ and $\sigma(J) \equiv 0$. Then if $(J^{\perp}/J,\rho_1,\sigma_1)$ is a kernel, so is (T,ρ,σ).

Proof. [P3, 3.5].

Consider the pair (R_m,τ_m) where

$$(R_m) = (E + \bar{F} \xrightarrow{\mu_m = \begin{pmatrix} m\alpha & 0 \\ 0 & m\bar{\alpha} \end{pmatrix}} F + \bar{E} \xrightarrow{\begin{pmatrix} j_m & 0 \\ 0 & k_m \end{pmatrix}} L_m + K_m$$

2.30
$$\tau_m = \begin{pmatrix} \lambda & \epsilon m^{-1}\bar{\alpha}^{-1} \\ m^{-1}\alpha^{-1} & 0 \end{pmatrix},$$

j_m and k_m are the cokernel projections of $m\alpha$ and $m\bar{\alpha}$, and $m \in \mathbb{Z}$ is chosen so large that the $(n\times n)$-matrix $m\lambda_s\alpha$ has entries in A. By (2.21) (R_m,τ_m) covers some ϵ-form (S_m,φ_m,ψ_m), which by (2.28) is hyperbolic. The following lemma together with (2.28) completes the proof of (2.22).

2.31. Lemma. Let the submodule J of $V + (L_m + K_m) = V + S_m$ be given by $J = j_m(\alpha(E))$. Then $J \subseteq J^{\perp}$, $\psi_m(J) \equiv 0$, and, setting $(T,\rho,\sigma) = (V,\zeta,\xi) \perp (S_m,\varphi_m,\psi_m)$, the induced $(J^{\perp}/J,\rho_1,\sigma_1)$ is isometric to $(V,\zeta,\xi) \perp (S,\varphi,\psi)$.

Proof. Since $J \subseteq L_m + K_m$, V is orthogonal to J so $V \subseteq J^{\perp}$. Thus there is a copy of (V,ζ,ξ) in $(J^{\perp}/J,\rho_1,\sigma_1)$, so it suffices to show that J, viewed as a submodule of $S_m = L_m + K_m$, satisfies the conditions $J \subset J^{\perp}$, $\psi_m(J) \equiv 0$, and $(J^{\perp}/J,\rho_1,\sigma_1) \cong (S,\varphi,\psi)$.

From the commutative diagram

$$\begin{array}{ccc} E & \xrightarrow{m} & E \\ \downarrow{\scriptstyle m\alpha} & & \downarrow{\scriptstyle \alpha} \\ F & \xrightarrow[=]{1} & F \\ \downarrow{\scriptstyle j_m} & & \downarrow{\scriptstyle j} \\ L_m & \dashrightarrow & L \end{array}$$

there is an induced surjection $L_m \to L$ with kernel $j_m(\alpha(E)) = J$. Hence

2.32 $$j_m(\alpha(E)) = J \rightarrowtail L_m \twoheadrightarrow L$$

is exact, where the first map is the inclusion. Similarly, the diagram

$$\begin{array}{ccc} \bar{F} & \xrightarrow[=]{1} & \bar{F} \\ \downarrow{\scriptstyle \bar{\alpha}} & & \downarrow{\scriptstyle m\bar{\alpha}} \\ \bar{E} & \xrightarrow{m} & \bar{E} \\ \downarrow{\scriptstyle k} & & \downarrow{\scriptstyle k_m} \\ K & \dashrightarrow & K_m \end{array}$$

yields an injection $K \to K_m$ whose image is $k_m(m\bar{E})$. K will be identified with $k_m(m\bar{E})$,

2.33 $$K \equiv k_m(m\bar{E}) \subseteq K_m.$$

To show that $J \subseteq J^\perp$ and $J^\perp/J \cong L + K = S$ it suffices by (2.32) to show that $J^\perp = L_m + K$; to show this it suffices to show that $L_m \subseteq J^\perp$ and $J^\perp \cap K_m = K$.

To see $L_m \subset J^\perp$, let $x \in J$, $y \in L_m$ and choose $\tilde{x},\tilde{y} \in F$ such that $j_m(\tilde{x}) = x$, $j_m(\tilde{y}) = y$; since $x \in J = j_m(\alpha(E))$, $\tilde{x} = \alpha(e)$ for some $e \in E$. (R_m,τ_m) covers (S_m,φ_m,ψ_m) so $\varphi_m(x,y) = r\tau_m(\tilde{x},\tilde{y})$, which equals $\lambda(\tilde{x},\tilde{y})$ using the upper left corner of τ_m. Hence $\varphi_m(x,y) = r\tau_m(\tilde{x},\tilde{y}) = r\lambda(\tilde{x},\tilde{y}) = r\lambda(\alpha(e),\tilde{y}) =$ (since τ covers (S,φ,ψ)) $\varphi(j\alpha(e),j(\tilde{y})) = 0$ ($j\alpha \equiv 0$). This shows $L_m \subset J^\perp$. To see that $J^\perp \cap K_m = K$ it suffices by (2.33) and (2.32) to show, given $f \in \bar{E}$, that if $\varphi_m(k_m(f), j_m\alpha(e)) = 0$ for all $e \in E$, then

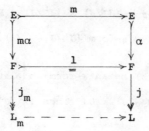

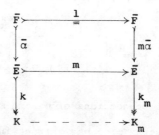

$f = mf'$, for some $f' \in \bar{E}$. Indeed, $\varphi_m(k_m(f), j_m\alpha(e)) = r\tau_m(f,\alpha(e)) = r(m^{-1}f(e))$ where the last equality follows using the lower left corner of τ_m and $f(e)$ means $f \in \bar{E} = \mathrm{Hom}(E,A)$ evaluated on $e \in E$. But if $r(m^{-1}f(e)) = 0$ for all $e \in E$, then $f(e) \in mA$ for all e, which implies f is divisible by m in $\bar{E}$ as required. As observed above this shows $J^{\perp}/J \cong S$.

The proof that $\psi_m(J) \equiv 0$ and $(J^{\perp}/J, \rho_1, \sigma_1) \cong (S, \varphi, \psi)$ uses the same kind of equations as above, together with (2.32), (2.33) and the fact that τ_m is roughly $m\tau$ (except for λ). This is left to the reader.

As mentioned above we should now proceed to a definition of equivalence classes of trivial structures on an object of $F_0^{\epsilon}(B/A)$ (a K_1-functor). In our case the two trivial structures are kernel structures on an object of $F_0^{\epsilon}(B/A)$ and are not quite arbitrary: one is required to be hyperbolic (see Remark (2.41)).

To begin with let $H, K \in \mathfrak{D}_F^1$, let $\mathscr{H}(H) = (H^{\wedge} + H, \varphi_h, \psi_h)$ be the hyperbolic ϵ-form associated to H, and let $\Delta\colon K \to H^{\wedge} + H$ be a map with $\Delta(K)$ totally isotropic ($\varphi_h|\Delta(K)\times\Delta(K) \equiv 0 \equiv \psi_h|\Delta(K)$). If $\Delta = (\gamma, \alpha)$, where $\gamma\colon K \to H^{\wedge}$ and $\alpha\colon K \to H$, there is a $(-\epsilon)$-hermitian form $\zeta\colon K \times K \to B/A$ defined by setting $\zeta(k,k') = \varphi_h(\gamma(k), \alpha(k'))$, for $k,k' \in K$. ζ is not in general nonsingular and by definition of φ_h, $\zeta(k,k') = \nu(\gamma(k), \alpha(k'))$, ν the natural form (2.6c). The $(-\epsilon)$-symmetry of ζ follows from the assumption that $\mathrm{im}(\Delta)$ is totally isotropic under φ_h.

2.34 <u>Definition</u>. Let $\epsilon = \pm 1$. An <u>ϵ-formation</u> is a 4-tuple (K,H,Δ,ξ) where K and $H \in \mathfrak{D}_F^1$, $\Delta\colon K \to H^{\wedge} + H$ is an injection whose image is a subkernel of $\mathscr{H}(H)$, and (K,ζ,ξ) is a $(-\epsilon)$-linking form (2.5d), where $\Delta = (\gamma,\alpha)$, $\zeta(k,k') = \nu(\gamma(k),\alpha(k'))$, ν is the natural form and $k,k' \in K$. An isomorphism between ϵ-formations is given by module isomorphisms of the K and H terms, preserving Δ and ξ. Let $F_1^{\epsilon}(B/A)$ denote the set of isomorphism classes of ϵ-formations.

2.35 Remarks. (a) $\mathrm{Ad}(\zeta) = \alpha^{\wedge}\gamma\colon K \to K^{\wedge}$. For if $k,k' \in K$, then $\{\alpha^{\wedge}\gamma(k)\}(k') = \nu(\alpha^{\wedge}\gamma(k),k') = \nu(\gamma(k),\alpha(k')) = \{\mathrm{Ad}(\zeta)(k)\}(k')$. (b) Definition (2.34) should be compared with the notion of formation in [Ra 1] or [Nov]. (c) The ξ-term in (K,H,Δ,ξ) is called for in the geometry of Chapter Seven. It is definitely extra structure, and should be compared to the splitting in Sharpe's definition of "split unitary group" [Sh 1]; in the last chapter we will show how ξ carries the Arf invariant.

We now define three operations on $F_1^{\epsilon}(B/A)$ leading to the definition in (2.38) of $L_1^{\epsilon}(B/A)$. In [P3,§2] there is a discussion of the analogy between the constructions in (2.36) and the coset relation on a unitary group relative to its commutator subgroup.

2.36 Proposition. Let $\theta_1 = (K_1,H_1,\Delta_1,\xi_1) \in F_1^{\epsilon}(B/A)$ be given, $\Delta_1 = (\gamma_1,\alpha_1)$, $\epsilon = \pm 1$.

a) Let (H_1,ρ,σ) be a $(-\epsilon)$-linking form and let $K_2 = K_1$, $H_2 = H_1$, $\alpha_2 = \alpha_1$, $\gamma_2 = \gamma_1 + \mathrm{Ad}(\rho)\cdot\alpha$ and $\xi_2 = \xi_1 + \sigma\cdot\alpha$. Then $(K_2,H_2,(\gamma_2,\alpha_2),\xi_2)$ is an ϵ-formation, denoted $\chi_{(H,\rho,\sigma)}\theta_1$.

b) Let $K_2 = K_1$, $H_2 = H_1^{\wedge}$, $(\gamma_2,\alpha_2) = (\epsilon\alpha_1,\gamma_1)$ and $\xi_2 = \epsilon\bar{\xi}_1$ (where $\bar{\xi}_1(x) = \overline{\xi_1(x)}$, $x \in K_1$). Then $(K_2,H_2,(\gamma_2,\alpha_2),\xi_2)$ is an ϵ-formation, denoted $\omega\theta_1$.

c) Let $E = (J \rightarrowtail H_2 \overset{p}{\twoheadrightarrow} H_1)$ be a short exact sequence of A-modules in $\mathfrak{D}_F^1$. Let $\mathcal{P}(p,H_1,\alpha_1)$ denote the pull back of p and α_1 in

2.37
$$\begin{array}{ccccc}
J & \rightarrowtail & \mathcal{P}(p,H_1,\alpha_1) & \overset{q}{\twoheadrightarrow} & K_1 \\
\Big\downarrow{=} & & \Big\downarrow & & \Big\downarrow{\alpha_1} \\
J & \rightarrowtail & H_2 & \overset{p}{\twoheadrightarrow} & H_1
\end{array}$$

Let $K_2 = \mathcal{P}(p,H_1,\alpha_1)$ and let α_2 be the induced map $\mathcal{P}(p,H_1,\alpha_1) \to H_2$; let $\gamma_2 = p^{\wedge}\cdot\gamma_1\cdot q$; and let $\xi_2 = \xi_1\cdot q$. Then $(K_2,H_2,(\gamma_2,\alpha_2),\xi_2)$ is an ϵ-formation, denoted $\sigma_E\theta_1$.

Proof. (a) and (c) are straightforward and left to the reader.

(b). The natural form $\nu_2\colon H \times H^\wedge = (H^\wedge)^\wedge \times H^\wedge \to B/A$ is defined by $\nu_2(h,f) = \overline{f(h)}$ where $f \in H^\wedge$, $h \in H$. Thus $\zeta_2\colon K_2 \times K_2 \to B/A$ is given by $\zeta_2(x,y) = \nu_2(\gamma_2(x),\alpha_2(y)) = \overline{\alpha_2(y)(\gamma_2(x))} = \overline{\gamma_1(y)(\epsilon\alpha_1(x))} = \epsilon\,\overline{\zeta_1(y,x)}$. From this it is easy to verify that (K_2,ζ_2,ξ_2) is a $(-\epsilon)$-linking form and that $\omega\theta_1$ is an ϵ-formation.

2.38 Definition. Let $\theta_i = (K_i,H_i,\Delta_i,\xi_i) \in F_1^\epsilon(B/A)$, $i = 1,2$. Set $\theta_1 \perp \theta_2 = (K_1+K_2,H_1+H_2,\Delta_1+\Delta_2,\xi_1+\xi_2)$. Let the zero formation, $\underline{0} \in F_1^\epsilon(B/A)$, the the 4-tuple with $H = (0) = K$. Viewing the constructions (2.36)(a)-(c) as operations on $F_1^\epsilon(B/A)$ let $L_1^\epsilon(B/A) = F_1^\epsilon(B/A)/\sim$, where "$\sim$" is the equivalence relation generated by these operations. (Thus $\theta_1 \sim \theta_2$ if and only if θ_1 can be converted to θ_2 by some sequence of operations in (2.36) and their inverses.)

2.39 Proposition. $L_1^\epsilon(B/A)$ is a commutative semigroup with addition induced by "$\perp$" and zero element represented by the zero formation.

Proof. This is obvious.

2.40 Remark. There is asymmetry in the definition of an ϵ-formation (K,H,Δ,ξ): while both K and H are subkernels, H is a summand of $H^\wedge + H$ while K need not be. For essentially this reason, algebraic construction of inverses in $L_1^\epsilon(B/A)$ is difficult. We will show geometrically in Chapter Seven that $L_1^\epsilon(B/A)$ is a group.

The asymmetry can be explained in the following way. Roughly speaking, objects of $F_1^\epsilon(B/A)$ compare two kernel structures on an element of $F_0^\epsilon(B/A)$. In our definitions one of these structures is hyperbolic; in general, we can modify one of the kernel structures to a hyperbolic structure by using Theorem (2.22) with $V = 0$. This modification can be formalized and turns out to be a form of stabilization (2.36(c)). In this way one can see that our definition of $L_1^\epsilon(B/A)$ and that of Karoubi

[K] are the same. His definition has better formal algebraic properties (for example the easy construction of inverses) while ours is easier to work with geometrically.

Next we list some simple properties of formations. The first is immediate from (2.36)(c) and the universal property of the pull-back construction.

2.41 <u>Proposition</u>. Let $\theta_i = (K_i, H_i, \Delta_i, \xi_i) \in F_1^{\epsilon}(B/A)$, $i = 1,2$, $\Delta_i = (\gamma_i, \alpha_i)$, and let $E = (J \rightarrowtail H_2 \overset{p}{\twoheadrightarrow} H_1)$ be a short exact sequence in $\mathfrak{D}_F^1$. Then $\theta_2 = \sigma_E \theta_1$ if and only if

a) there is a short exact sequence $L \rightarrowtail K_2 \overset{q}{\twoheadrightarrow} K_1$ of A-modules in $\mathfrak{D}_F^1$ and a commutative diagram

2.42 <u>Diagram.</u>

$$\begin{array}{ccccc} L & \rightarrowtail & K_2 & \overset{q}{\twoheadrightarrow} & K_1 \\ \cong \downarrow \alpha_2|L & & \downarrow \alpha_2 & & \downarrow \alpha_1 \\ J & \rightarrowtail & H_2 & \overset{p}{\twoheadrightarrow} & H_1 \end{array},$$

b) $\gamma_2 = p^{\wedge} \cdot \gamma_1 \cdot q$, and

c) $\xi_2 = \xi_1 \cdot q$.

2.43 <u>Proposition</u>. Let $\theta, \theta' \in F_1^{\epsilon}(B/A)$ satisfy $\sigma_E \theta = \sigma_E \theta'$, where E is a short exact sequence in $\mathfrak{D}_F^1$. Then $\theta = \theta'$ in $F_1^{\epsilon}(B/A)$.

<u>Proof</u>. It suffices to observe that θ_2 determines θ_1 in (2.41): α_1 is a quotient of α_2; γ_2 determines γ_1 since q is surjective and $p^{\wedge}$ is injective; and ξ_2 determines ξ_1 since q is surjective.

2.44 <u>Definition</u>. Let <u>$M_n(A,B)^{\times}$</u> denote the set of (n×n)-matrices over A which are invertible over B. If $\nu, \mu \in M_n(A,B)^{\times}$, the exact sequence [Ma, p. 51, Ex. 6]

$$\operatorname{cok} \nu \rightarrowtail \operatorname{cok}(\mu\nu) \twoheadrightarrow \operatorname{cok} \mu$$

is denoted <u>$E(\nu,\mu)$</u>. An isomorphism $E_1 \cong E_2$ between exact sequences

$E_i = (J_i \rightarrowtail I_i \twoheadrightarrow K_i)$ in $\mathfrak{D}^1_F$, $i = 1,2$, is given by isomorphisms of corresponding terms making the obvious diagram commute. An isomorphism $E_1 \cong E_2$ is said to be rel K (resp. I;J) if $K_1 = K_2$ (resp. $I_1 = I_2$; $J_1 = J_2$) and the isomorphism contains the identity map on K (resp. $I;J$). Finally if μ is an $(n\times n)$-matrix $\mu \perp I_m$ denotes the $(n+m) \times (n+m)$-matrix $\begin{pmatrix}\mu & 0\\ 0 & I_m\end{pmatrix}$.

2.45 Proposition. (a) If $\mu \in M_n(A,B)^{\times}$ and $E = (J \overset{i}{\rightarrowtail} I \twoheadrightarrow \operatorname{cok}\mu)$ is a short exact sequence in $\mathfrak{D}^1_F$, there exists $\nu \in M_k(A,B)^{\times}$ such that $E \cong E(\nu,\mu \perp I_m)$ (rel. $\operatorname{cok}\mu \equiv \operatorname{cok}(\mu \perp I_m)$), where $k = n + m$.

(b) $\eta \in M_n(A,B)^{\times}$ and $E = (J \overset{i}{\rightarrowtail} \operatorname{cok}\eta \twoheadrightarrow K)$ is a short exact sequence in $\mathfrak{D}^1_F$, then there exist $\mu,\nu \in M_k(A,B)^{\times}$ such that $\mu\nu = \eta \perp I_m$, $E \cong E(\nu,\mu)$ (rel. $\operatorname{cok}\eta \equiv \operatorname{cok}(\eta \perp I_m)$), and $k = n + m$.

Proof. (a) Let $E \overset{\mu}{\rightarrowtail} F \twoheadrightarrow \operatorname{cok}\mu$ be a free resolution of $\operatorname{cok}\mu$ and $C \overset{\nu'}{\rightarrowtail} D \twoheadrightarrow J$, one of J. The standard construction [CE,p. 79] for an induced resolution of I fits into the commutative diagram with short exact columns:

$$\begin{array}{ccccc}
C & \longrightarrow & C + E & \longrightarrow & E\\
\big\downarrow \nu' & & \big\downarrow \begin{pmatrix}\nu' & \varkappa\\ 0 & \mu\end{pmatrix} & & \big\downarrow \mu\\
D & \longrightarrow & D + F & \longrightarrow & F\\
\big\downarrow & & \big\downarrow & & \big\downarrow\\
J & \rightarrowtail & I & \twoheadrightarrow & \operatorname{cok}\mu,
\end{array}$$

2.46 Figure.

where $\varkappa: E \to D$. Setting $\nu = \begin{pmatrix}\nu' & \varkappa\\ 0 & I\end{pmatrix}$ the matrix equation $\begin{pmatrix}\nu' & \varkappa\\ 0 & \mu\end{pmatrix} = \begin{pmatrix}I & 0\\ 0 & \mu\end{pmatrix}\nu$ yields the conclusion.

(b). Form the pull back diagram

$$\begin{array}{ccc}
E & \overset{=}{\longrightarrow} & E\\
\big\downarrow & & \big\downarrow \eta\\
\theta = \theta(i,\operatorname{cok}\eta,j) & \longrightarrow & F\\
\big\downarrow & & \big\downarrow j\\
J & \longrightarrow & \operatorname{cok}\eta
\end{array}$$

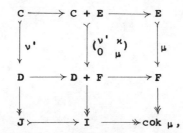

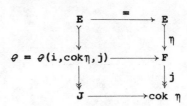

where E and F are A-free of rank, say, n. $\mathcal{P}$ is clearly $\mathbb{Z}$-torsion-free and Schanuel's Lemma [Ma, p. 101] shows it is stably free. Thus, taking the direct sum of the upper part of the diagram with

$$\begin{array}{ccc} A^m & \xrightarrow{=} & A^m \\ {\scriptstyle =}\downarrow & & \downarrow{\scriptstyle =} \\ A^m & \xrightarrow{=} & A^m \end{array}$$

for m sufficiently large completes the construction, where ν is taken to be the induced map $E + A^n \to \mathcal{P} + A^n$ and μ the map $\mathcal{P} + A^n \to F + A^n$.

The last property of formations follows easily from the definition of "pull-back."

2.47 <u>Proposition</u>. Let $\mu,\nu,\rho \in M_n(A,B)^\times$ and let $\theta = (K, \operatorname{cok}\mu, \Delta, \xi) \in F_1^\epsilon(B/A)$. Then $\sigma_{E(\rho,\mu\nu)}(\sigma_{E(\nu,\mu)}\theta) = \sigma_{E(\nu\rho,\mu)}\theta$.

Chapter Three: Moore Spaces

In this chapter we study what we call "Moore spaces", and list some properties to show they behave like mod t spheres. However, we cannot refer directly to existing treatments of mod t homotopy theory, because our Moore spaces have nontrivial fundamental group, we use specific models for them in later chapters, and because we need a slightly more general space (see Def. (3.1)). The main results of this chapter are the universal coefficient theorem (3.14) and the consequent representability of homology classes in highly-connected surgery problems by maps of Moore spaces ((3.17) and (3.18)).

3.1 Definition. Let $k, t_1, \ldots, t_n$ be positive integers. Let S^k be an oriented k-sphere, remove from it the interiors of $\Sigma_{i=1}^n t_i$ k-discs $D_1^k, \ldots, D_{t_1}^k, D_{t_1+1}^k, \ldots$, and identify the first t_1 boundary components $\partial D_1^k \approx \cdots \approx \partial D_{t_1}^k$ in orientation-preserving fashion, do the same for the next t_2 ∂D^k's, and so on. The resultant space is called a generalized Moore space of order $(t_1, \ldots, t_n)$ and dimension k, denoted $\mathfrak{M}^k_{\{t_1,\ldots,t_n\}}$. After identification the Σt_i (k-1)-spheres, ∂D_i^k, are converted to n disjoint (k-1)-spheres the union of which is the bockstein $b\mathfrak{M}^k_{\{t_1,\ldots,t_n\}}$. The interior $\mathring{\mathfrak{M}}^k_{\{t_1,\ldots,t_n\}}$ is $\mathfrak{M}^k_{\{t_1,\ldots,t_n\}} - b\mathfrak{M}^k_{\{t_1,\ldots,t_n\}}$. In case $n = 1$, $\mathfrak{M}^k_t$ is called a Moore space.

The case n = 1 in the following proposition explains the use of the term "Moore space."

3.2 Proposition. If $m\colon \mathbb{Z} \to \mathbb{Z}^n$ is given by $m(1) = (t_1, \ldots, t_n)$, $t_i > 0$, then if $k \geq 3$ and $p = (\Sigma t_i) - n$, we have $\pi_1 \mathfrak{M}^k_{\{t_1,\ldots,t_n\}} = \mathbb{Z} * \cdots * \mathbb{Z}$, the p-fold free product, $H_i \mathfrak{M}^k_{\{t_1,\ldots,t_n\}} = 0$ for $i > 1$, $i \neq k-1$, and $H_{k-1}\mathfrak{M}^k_{\{t_1,\ldots,t_n\}} = \operatorname{cok}(m)$.

<u>Proof</u>. Let $\mathfrak{M} = \mathfrak{M}^k_{\{t_1,\ldots,t_n\}}$. If D is a k-disc in $\mathring{\mathfrak{M}}$, $\mathring{\mathfrak{M}} - D$ has the homotopy type of (p circles) $\vee$ (n S^{k-1}'s); this gives the π_1 statement. To compute homology, use the Mayer-Vietoris sequence

$$\cdots \longrightarrow H_{i+1}(\mathfrak{M}) \longrightarrow H_i(\partial D) \longrightarrow H_i(D) + H_i(\mathfrak{M} - \mathring{D}) \longrightarrow H_i(\mathfrak{M}) \longrightarrow \cdots$$

together with the fact that $[\partial D]$ (= generator) $\in H_{k-1}(\partial D)$ is homologous in $\mathfrak{M} - \mathring{D}$ to $(t_1,\ldots,t_n) \in \mathbb{Z}^n = H_{k-1}(\mathfrak{M} - \mathring{D})$.

3.3 <u>Corollary</u>. If $\mathfrak{M} = \mathfrak{M}^k_{\{t_1,\ldots,t_n\}}$ and $k \geq 3$, then $H^k(\mathfrak{M}, b\mathfrak{M}) = \mathbb{Z}$ and $\delta\colon H^{k-1}(b\mathfrak{M}) \to H^k(\mathfrak{M}, b\mathfrak{M})$ may be identified with $m^*\colon \mathbb{Z}^n \to \mathbb{Z}$, $m^*(g_i) = t_i \in \mathbb{Z}$, where $\{g_1,\ldots,g_n\}$ is a $\mathbb{Z}$-basis of $\mathbb{Z}^n$.

<u>Proof</u>. Consider the diagram of exact sequences

3.4 <u>Diagram</u>

$$\begin{array}{ccccc}
H_k(\mathfrak{M}, b\mathfrak{M}) & \xrightarrow{\ \partial\ } & H_{k-1}(b\mathfrak{M}) & \longrightarrow & H_{k-1}(\mathfrak{M}) \\
\downarrow & & \downarrow \cong & & \downarrow = \\
H_k(\mathfrak{M}, \mathfrak{M}-\mathring{D}) & \longrightarrow & H_{k-1}(\mathfrak{M}-\mathring{D}) & \longrightarrow & H_{k-1}(\mathfrak{M})
\end{array}$$

where $D \subset \mathring{\mathfrak{M}}$ is a k-disc. Since (as in the proof of (3.2)) the generator of $H_{k-1}(\partial D) = H_k(D, \partial D) = H_k(\mathfrak{M}, \mathfrak{M} - \mathring{D})$ goes to $(t_1,\ldots,t_n) \in H_{k-1}(\mathfrak{M} - \mathring{D})$, the generator of $H_k(\mathfrak{M}, b\mathfrak{M})$ behaves the same under ∂. Dualizing to cohomology finishes the proof.

3.5 <u>Proposition</u>. Suppose given $\alpha_i \in \pi_{k-1}(X)$, $1 \leq i \leq n$, with $\Sigma\, t_i\alpha_i = 0$, $t_i \in \mathbb{Z}^+$, $k \geq 3$, and X a simply-connected space. Then there is a map $b\colon \mathfrak{M}^k_{\{t_1,\ldots,t_n\}} \to X$ extending $\dot{\cup}\, a_i\colon \dot{\cup}\, S^{k-1} \to X$, where $\dot{\cup}\, S^{k-1}$ is the disjoint union of n (k-1)-spheres, identified with $b\mathfrak{M}^k_{\{t_1,\ldots,t_n\}}$, and $[a_i] = \alpha_i \in \pi_{k-1}(X)$.

<u>Proof</u>. Denote $\mathfrak{M} = \mathfrak{M}^k_{\{t_1,\ldots,t_n\}}$. By obstruction theory and (3.2), (3.3), there is a single obstruction $\theta \in H^k(\mathfrak{M}, \dot{\cup}\, S^{k-1}; \pi_{k-1}(\vee S^{k-1}))$ to extending the map $\dot{\cup}\, S^{k-1} \to \vee S^{k-1}$, which is the identity on each term,

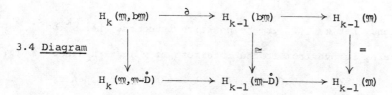

to a map $\mathfrak{M} \to \vee S^{k-1}$. By [Sp, Thm. 8.17], $\theta = \delta i$, where $i \in H^{k-1}(\dot{\cup}\, S^{k-1}; \pi_{k-1}(\vee S^{k-1}))$ is the canonical element, and $\delta\colon H^{k-1}(\dot{\cup}\, S^{k-1}; \pi_{k-1}(\vee S^{k-1})) \to H^k(\mathfrak{M}, \dot{\cup}\, S^{k-1}; \pi_{k-1}(\vee S^{k-1}))$.

Hence the image of θ under the map is cohomology induced by the coefficient map $(\vee a_i)_\#\colon \pi_{k-1}(\vee S^{k-1}) \to \pi_{k-1}(X)$ is $\Sigma\, t_i \alpha_i$ by (3.3); this is the obstruction to constructing b and is zero by assumption.

3.6 Remark. b is not necessarily unique up to homotopy (3.14).

3.7 Definition. Let $k, t_1, \ldots, t_n$ be positive integers. Let D^{k+1} be an oriented $(k+1)$-disc, and let $D_1^k, \ldots, D_{t_1}^k, D_{t_1+1}^k, \ldots$ be $\Sigma\, t_i$ disjoint k-discs in ∂D^{k+1}. Then the relative generalized Moore space of order $(t_1, \ldots, t_n)$ and dimension $k+1$, $\bar{\mathfrak{M}}^{k+1}_{\{t_1,\ldots,t_n\}}$, is obtained from D^{k+1} by identifying the first t_1 k-discs to one another by orientation-preserving homeomorphisms, then identifying the next t_2 discs to one another, and so on, until the last t_n discs are identified to one another. The image under identification of the $\Sigma\, t_i$ k-discs is the bockstein, $b\bar{\mathfrak{M}}^{k+1}_{\{t_1,\ldots,t_n\}}$, the disjoint union of n k-discs. The interior $\mathring{\bar{\mathfrak{M}}}^{k+1}_{\{t_1,\ldots,t_n\}}$ is $\bar{\mathfrak{M}}^{k+1}_{\{t_1,\ldots,t_n\}} - b\bar{\mathfrak{M}}^{k+1}_{\{t_1,\ldots,t_n\}}$. If $n = 1$, $\bar{\mathfrak{M}}^{k+1}_t$ is a relative Moore space.

3.8 Proposition. A map $f\colon \mathfrak{M}^k_t \to X$, with $f_\#\colon \pi_1\mathfrak{M}^k_t \to \pi_1 X$ the zero map and $k \geq 3$, is null-homotopic if and only if it admits an extension over $\bar{\mathfrak{M}}^{k+1}_t$.

Proof. Let $\mathfrak{M} = \mathfrak{M}^k_t$. If $F_1\colon \mathfrak{M} \times I \to X$ is a null-homotopy of f, let $G\colon D^k \times I$ extend $F_1 \mid b\mathfrak{M} \times I$ with $G \mid D^k \times \{1\}$ constant and adjoin G to F_1 along $F_1 \mid b\mathfrak{M} \times I$. This gives a map $F_2\colon \bar{\mathfrak{M}} - \mathring{D} \to X$, $\mathring{D} \subset \mathring{\bar{\mathfrak{M}}}$ a k-disc, with $F_2 \mid \partial D$ constant. The map $F\colon \bar{\mathfrak{M}} \to X$ with $F \mid \bar{\mathfrak{M}} - \mathring{D} = F_2$ and $F|D$ constant is the required extension of f. The converse is similar.

3.9 Proposition. Let $k \geq 3$ and let $\mathfrak{M} = \mathfrak{M}^k_{\{t_1,\ldots,t_n\}}$,

$\bar{\mathfrak{M}} = \bar{\mathfrak{M}}^{k+1}_{\{t_1,\ldots,t_n\}}$. Then $\bar{\mathfrak{M}}$ has the homotopy type of a bouquet of circles, $\pi_1\mathfrak{M} \xrightarrow{\cong} \pi_1\bar{\mathfrak{M}}$, and $H_i(\bar{\mathfrak{M}},\mathfrak{M}) = \mathrm{cok}(m)$ if $i = k$ and is zero otherwise, where m is the map of (3.2).

<u>Proof</u>. Part (a) is obvious, (b) is a consequence of (a), and (c) follows from (a) and (3.2).

Here is the relative version of (3.5).

3.10 <u>Proposition</u>. Let $\alpha_i \in \pi_k(Y,X)$, $1 \leq i \leq n$, with $\Sigma\, t_i\alpha_i = 0$, $t_i \in \mathbb{Z}$, $k \geq 3$, and X and Y simply-connected. Then there is a map $b\colon (\bar{\mathfrak{M}}^{k+1}_{\{t_1,\ldots,t_n\}}, \mathfrak{M}^k_{\{t_1,\ldots,t_n\}}) \to (Y,X)$ of generalized Moore spaces extending $\dot{\cup}\, a_i\colon \dot{\cup}(D^k,S^{k-1}) \to (Y,X)$, where $\dot{\cup}(D^k,S^{k-1})$ is a disjoint union identified with $(b\bar{\mathfrak{M}}^{k+1}_{\{t_1,\ldots,t_n\}}, b\mathfrak{M}^k_{\{t_1,\ldots,t_n\}})$ and $[a_i] = \alpha_i = \pi_k(Y,X)$.

<u>Proof</u>. Compare (3.8) and (3.5).

Our goal is a "universal coefficient sequence" for homotopy groups. We begin with a geometric model for the addition of maps $f,g\colon (\bar{\mathfrak{M}}^{k+1}_t, \mathfrak{M}^k_t) \to (Y,X)$, which will be needed in Chapter Seven.

3.11 Let $I = [0,1]$. Write $I^{k+1} = I^k \times I$ and define $J = I^k \times 0 \cup \dot{I}^k \times I$, where $\dot{I}^k$ is the boundary of I^k. Let $p_n = n/2t + 1 \in I$, $1 \leq n \leq 2t + 1$ and set $I_m = [p_{2m-1}, p_{2m}]$, $m = 1,\ldots,t$. Let $D_m = I^2 \times (I\ )^{k-2} \times 1 \subset I^k \times 1 \subset I^{k+1}$. For $t = 2$, $k = 3$, $I^k \times 1$ is

3.11(a)<u>Figure</u>

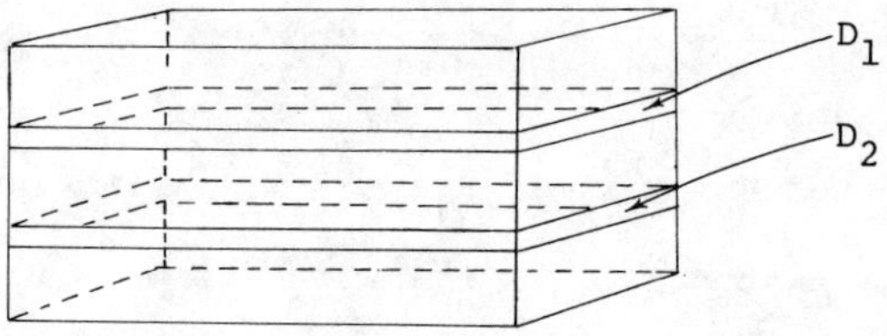

For each m, $D_m \subset \dot{I}^{k+1}$; we identify maps $g\colon (\bar{\mathfrak{M}}^{k+1}_t, \mathfrak{M}^k_t) \to (Y,X)$ with maps $\tilde{g}\colon (I^{k+1}, \dot{I}^{k+1} - \cup\, \mathring{D}_i) \to (Y,X)$ satisfying $\tilde{g} \mid D_i = \tilde{g} \mid D_j$, $1 \leq i, j \leq t$. Also, we want to assume g sends a subset of $\bar{\mathfrak{M}}$ carrying π_1 to a point x_0 of X (3.12). Let $J_m = I^2 \times (I_m)^{k-3} \times p_{2m-1} \times 1 \cup \partial(I^2 \times (I_m)^{k-3}) \times I_m \times 1$; this is a $(k-1)$-disc in ∂D_m (analogous to J in I^{k+1}) and

$\partial D_m - \mathring{J}_m = I^2 \times (I_m)^{k-3} \times P_{2m} \times 1$. Let K be the image of $J \cup (\cup J_i)$ under the identification to form $\mathfrak{M}_t^{k+1}$; then K carries $\pi_1 \mathfrak{M}_t^{k+1}$.

3.12 Lemma. Let g, $\tilde{g}$, Y, X be as above with $\pi_1 X = \pi_1 Y = 0$ and let $k \geq 3$. Then $\tilde{g}$ is homotopic as a map of pairs to $\tilde{g}'$ where $\tilde{g}'(J \cup (\cup J_i)) = x_0 \in X$; if $\tilde{g}_s$ is the homotopy, $0 \leq s \leq 1$, then $\tilde{g}_s|D_i = \tilde{g}_s|D_j$, $1 \leq i,j \leq t$.

Proof. Left to the reader.

3.13 Definition. Let K be as above and $k \geq 3$. The homotopy set $[\mathfrak{M}_t^{k+1}, \mathfrak{M}_t^k, K; Y, X, x_0]$ is denoted $\pi_{k+1}(Y,X;\mathbb{Z}_t)$.

3.14 Proposition. If $k \geq 3$, and X and Y are simply-connected, there is an exact sequence of abelian groups

$$\pi_{k+1}(Y,X) \otimes \mathbb{Z}_t \xrightarrow{p} \pi_{k+1}(Y,X;\mathbb{Z}_t) \xrightarrow{q} \mathrm{Tor}(\pi_k(Y,X), \mathbb{Z}_t).$$

Proof. To define a group structure on $\pi_{k+1}(Y,X;\mathbb{Z}_t)$, let $g_1, g_2\colon (\mathfrak{M}_t^{k+1}, \mathfrak{M}_t^k, K) \to (Y,X,x_0)$ be identified with $\tilde{g}_1, \tilde{g}_2\colon (I^{k+1}, \dot{I}^{k+1} - \cup \mathring{D}_i, J \cup (\bigcup_i J_i)) \to (Y,X,x_0)$, $\tilde{g}_n|D_i = \tilde{g}_n|D_j$, $n = 1,2$; $1 \leq i,j \leq t$. Define $\tilde{g}_1 * \tilde{g}_2\colon I^{k+1} \to Y$ by

$$\tilde{g}_1 * \tilde{g}_2(t_1,\ldots,t_{k+1}) = \begin{cases} \tilde{g}_1(2t_1, t_2, \ldots, t_{k+1}), & 0 \leq t_1 \leq \frac{1}{2} \\ \tilde{g}_2(2t_1 - 1, t_2, \ldots, t_{k+1}), & \frac{1}{2} \leq t_1 \leq 1 . \end{cases}$$

This is a well-defined map taking $J \cup (\cup J_i)$ to x_0 and passing under identification to a map $g_1 * g_2\colon (\mathfrak{M}_t^{k+1}, \mathfrak{M}_t^k, K) \to (Y,X,x_0)$. Following [Wh] it is easy to verify that "*" gives $\pi_{k+1}(Y,X;\mathbb{Z}_t)$ the structure of abelian group.

To define p let $h\colon (I^{k+1}, \dot{I}^{k+1}, J) \to (Y,X,x_0)$ represent $\alpha \in \pi_{k+1}(Y,X,x_0)$. Deform h to h' where $h'(\cup D_i) = x_0$. Then $h'(J \cup (\cup J_i)) = x_0$ and $h'|D_i = h'|D_j$, $1 \leq i,j \leq t$; h' thus defines an element of $\pi_{k+1}(Y,X,x_0;\mathbb{Z}_t)$. The map q is defined by restricting $h: (I^{k+1}, \dot{I}^{k+1} - \cup \dot{D}_i, J \cup (\cup J_i)) \to (Y,X,x_0)$ representing $\beta \in \pi_{k+1}(Y,X,x_0)$ to any D_m; (3.10) and (3.12) show q is surjective. Further details are left to the reader.

The following is a standard result about the way the universal coefficient sequences (3.14) fit together for large t.

3.15 Proposition. Let (Y,X) be a pair of spaces such that $\pi_1 X \cong \pi_1 Y = 0$, $\pi_k(Y,X)$ has exponent t, and $k \geq 3$. Then for any $s \in \mathbb{Z}$, $s \geq 1$, there is a map $r_s: \pi_{k+1}(Y,X;\mathbb{Z}_t) \to \pi_{k+1}(Y,X;\mathbb{Z}_{ts})$ and a commutative exact diagram

$$\begin{array}{ccccc} \pi_{k+1}(Y,X) \otimes \mathbb{Z}_t & \xrightarrow{p_t} & \pi_{k+1}(Y,X;\mathbb{Z}_t) & \xrightarrow{q_t} & \pi_k(Y,X) \\ \downarrow m_s & & \downarrow r_s & & \downarrow = \\ \pi_{k+1}(Y,X) \otimes \mathbb{Z}_{ts} & \xrightarrow{p_{ts}} & \pi_{k+1}(Y,X;\mathbb{Z}_{ts}) & \xrightarrow{q_{ts}} & \pi_k(Y,X) \end{array}$$

where m_s is induced by the natural injection $\mathbb{Z}_t \to \mathbb{Z}_{ts}$.

Proof. Let $[g] \in \pi_{k+1}(Y,X;\mathbb{Z}_t)$ be given, where $g: (\bar{\mathfrak{M}}_t^{k+1}, \mathfrak{M}_t^k, K) \to (Y,X,x_0)$ is represented as in (3.11) by $\tilde{g}: (I^{k+1}, \dot{I}^{k+1} - \bigcup_i D_i, J \cup (\bigcup_i J_i)) \to (Y,X,x_0)$. Define $\tilde{g}_s: I^{k+1} \to Y$ by $\tilde{g}_s(t_1,\ldots,t_{k+1}) = \tilde{g}(t_1,t_2,st_3-n,t_4,\ldots,t_{k+1})$, $\frac{n}{s} \leq t_3 \leq \frac{n+1}{s}$, $n = 0,1,\ldots,s-1$ (compare Fig. (3.11)(a))

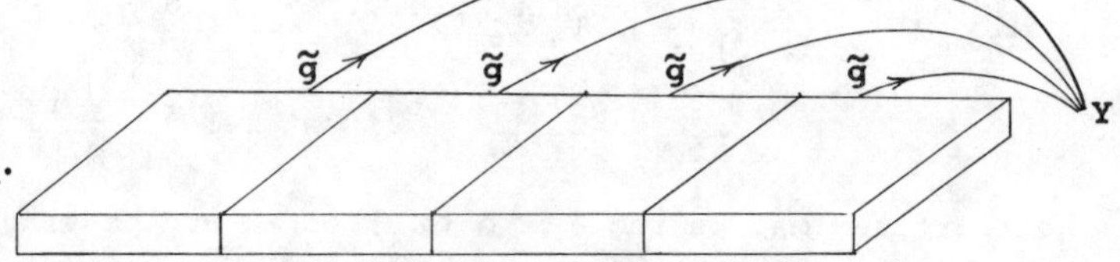

3.16 Figure.

It is easy to verify that (in the manner of (3.11)) $\tilde{g}_s$ induces $g_s: (\bar{\mathfrak{M}}_{ts}^{k+1}, \mathfrak{M}_{ts}^k, K) \to (Y,X,x_0)$, that $r_s[g] = [g_s]$ is well defined, and that the required diagram commutes.

Interpreting the homotopy groups of a map as relative homotopy groups in the usual way [Hi] and using the Hurewicz theorem, we obtain the following result as a corollary of the surjectivity of q in (3.14).

3.17 <u>Proposition</u>. Let $g: N \to Y$ be a map of simply-connected spaces such that $\pi_i(g) = 0$, $i < k - 1$, $k \geq 3$ and $tH_k(g) = 0$, for some $t \in \mathbb{Z}^+$. Then for each $x \in H_k(g)$ there is a commutative square

$$\underline{\underline{\beta}} = \begin{array}{ccc} \mathfrak{M}_t^k & \longrightarrow & N \\ \downarrow & & \downarrow g \\ \bar{\mathfrak{M}}_t^{k+1} & \longrightarrow & Y \end{array}$$

such that $hq[\underline{\underline{\beta}}] = x$ where $q: \pi_{k+1}(g;\mathbb{Z}_t) \to \pi_k(g)$ is given in (3.14) and $h: \pi_k(g) \to H_k(g)$ is the Hurewicz homomorphism.

3.18. Relativizing further, for any quadrad of maps and spaces

$$\Phi = \begin{array}{ccc} M & \longrightarrow & X \\ \downarrow & & \downarrow \\ N & \longrightarrow & Y \end{array}$$

there is defined [Hi] a homotopy set $\pi_k(\Phi)$ which is an abelian group if $k \geq 3$. Now the union of two relative generalized Moore spaces along their "boundaries" is a generalized Moore space: $(\bar{\mathfrak{M}}^{k+1}_{\{t_1,\ldots,t_n\}})_+ \cup (\bar{\mathfrak{M}}^{k+1}_{\{t_1,\ldots,t_n\}})_- = \mathfrak{M}^{k+1}_{\{t_1,\ldots,t_n\}} \subseteq \bar{\mathfrak{M}}^{k+2}_{\{t_1,\ldots,t_n\}}$, where $(\bar{\mathfrak{M}}^{k+1}_{\{t_1,\ldots,t_n\}})_+ \cap (\bar{\mathfrak{M}}^{k+1}_{\{t_1,\ldots,t_n\}})_- = \mathfrak{M}^{k}_{\{t_1,\ldots,t_n\}}$. In (4B.26) we will use the obvious generalization of (3.10) involving maps of the quadrad of inclusions

$$\text{3.19} \qquad \begin{array}{ccc} \mathfrak{M}^{k}_{\{t_1,\ldots,t_n\}} & \longrightarrow & (\bar{\mathfrak{M}}^{k+1}_{\{t_1,\ldots,t_n\}})_+ \\ \downarrow & & \downarrow \\ (\bar{\mathfrak{M}}^{k+1}_{\{t_1,\ldots,t_n\}})_- & \longrightarrow & \bar{\mathfrak{M}}^{k+2}_{\{t_1,\ldots,t_n\}} \end{array}$$

into Φ; where $n = 1$ and $t_1 = t$, the set of homotopy classes of such maps, $\pi_{k+2}(\Phi;\mathbb{Z}_t)$, may be studied as above. In particular, using the Hurewicz theorem for quadrads [Na] and the generalization of (3.14) to homotopy groups of quadrads, the reader may easily formulate a version of (3.17) for quadrads $\mathfrak{g}$ in place of g.

3.20 Remark. The group $\pi_k(X,x_0;\mathbb{Z}_t)$ is defined to be $[\mathfrak{M}_t^k,K;X,x_0]$ $(k \geq 3)$. Let $E = \{(t_1,\ldots,t_{k+1}) \in \dot{I}^{k+1} \mid t_1 = 1\}$ and let $\bar{E}$ be its image in K. Then $\bar{E} \approx \bar{\mathfrak{M}}_t^k$ and we may take $\bar{E}$ to be one of the terms in the decomposition discussed in (3.18),

$$\mathfrak{M}_t^k = (\bar{\mathfrak{M}}_t^k)_+ \cup (\bar{\mathfrak{M}}_t^k)_-$$

where $(\bar{\mathfrak{M}}_t^k)_+ \cap (\bar{\mathfrak{M}}_t^k)_- = \mathfrak{M}_t^{k+1}$. When X is a manifold and $f\colon \mathfrak{M}_t^k \to X$ is an imbedding (in Chapter Four), x_0 will be replaced by a small disc $D \supseteq \{x_0\}$ so that f imbeds K, and hence $\bar{E} \approx (\bar{\mathfrak{M}}_t^k)_+$ into D.

Chapter Four: A Class of Stratified Spaces and Some of its Geometric Properties.

The geometric content of this paper is a "local" theory of surgery. In a formal geometric sense, local surgery differs from ordinary surgery only in that surgery is done on a certain class of stratified spaces, instead of spheres. Chapter Four is devoted to studying the geometric properties of these spaces in order to carry out the surgery in Chapters Five, Six and Seven. It turns out that most of these properties can be formulated outside the context of surgery theory without extra effort. From this it is seen that these stratified spaces can be generalized and applied elsewhere.

There are three parts to Chapter Four: A, B and C. Part A studies Moore spaces, part B the more general conglomerate Moore spaces, and part C a construction needed in Chapter Seven. A more detailed account of Chapter Four is found at the beginning of each of its parts. We work exclusively in the piecewise-linear category for consistency and because a technical device (the "collar" in (4.A.1)) is needed to state framing conditions. All results remain valid in the smooth and topological categories.

Conventions: Throughout Chapter Four, the letter "k" will denote an integer ≥ 3:

$$\boxed{k \geq 3}$$

unless otherwise stated. This hypothesis is needed so often (and is sufficient for applications) that it is stated in advance. It is always assumed that the fundamental groups of manifolds are finite; this is for convenience and because it is what is needed in later chapters. In fact, this hypothesis is rarely necessary for the results of this chapter to be valid. We always assume (as in Chapter 3) that a map $F: \mathfrak{M}_t^k \to X$ induces the trivial homomorphism on π_1.

Chapter 4A: Geometry of Moore Spaces

Chapter 4A begins with a slight modification of the definition of a Moore space given in (3.1): instead of identifying the t boundary components of $S^k - (\bigcup_1^t \mathring{D}_i^k)$, we identify collar neighborhoods of the boundary components in orientation-preserving fashion. With this definition a neighborhood of $x \in \mathfrak{M}_t^k$ looks like R^k or $R^{k-1} \times C(t)$ (the latter when $x \in b\mathfrak{M}_t^k$, the bockstein) where $C(t)$ is the union at the cone point of the cone on t points with an interval:

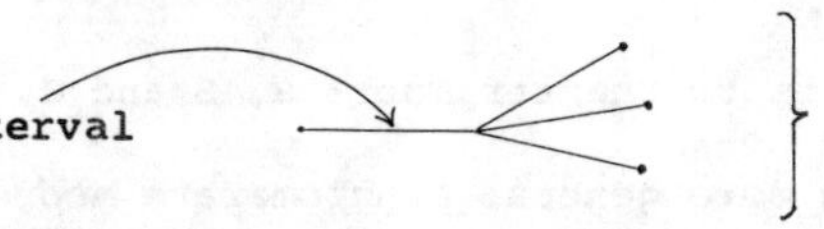

The interval is a piecewise-linear version of the 1-frame field on $b\mathfrak{M}_t^k$ given by the tangent direction of the sheets coming together at bockstein in a <u>smooth</u> model for $\mathfrak{M}_t^k$:

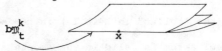

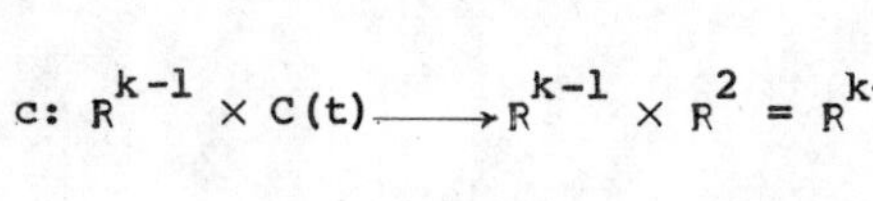

Next we construct a specific imbedding $c: C(t) \to R^2$ and a map $F: \mathfrak{M}_t^k \to R^{k+1}$ which is an imbedding in a neighborhood U of the bockstein, has the local form

$$(id) \times c: R^{k-1} \times C(t) \longrightarrow R^{k-1} \times R^2 = R^{k+1}$$

near points $x \in b\mathfrak{M}_t^k$, and is an immersion restricted to $\mathfrak{M}_t^k - b\mathfrak{M}_t^k$ (a manifold). The normal direction to $c(C(t))$ in R^2 defines a one-frame field ϵ on U which extends to $\mathfrak{M}_t^k$, is transverse to $F(\mathfrak{M}_t^k) \subseteq R^{k+1}$, and is called the "distinguished field". The "total space" $n\mathfrak{M}_t^k$ of ϵ is a canonical codimension-one thickening of $\mathfrak{M}_t^k$, immersed in R^{k+1} by the obvious extension of F, and is thus a parallelized manifold.

An immersion (4A.15) $F: \mathfrak{M}_t^k \to M$, where M is a piecewise-linear manifold, is defined to be a map extending to $F': n\mathfrak{M}_t^k \to M$. From this follows "immersion theory" ([HP]) for Moore spaces and the consequent representability of homology classes in the kernel of a normal map by framed immersions of Moore spaces (4A.21). Next it is shown (4A.29) that the intersection numbers of immersed k-dimensional Moore spaces $\mathfrak{M}_t^k$ in a 2k-manifold are invariants mod t of regular homotopy, and that a canonical "replication" process (4A.35) allows all intersections to be removed, without changing bocksteins (4A.36). Finally, linking numbers are studied and it is shown how they determine the intersection of an immersed Moore space with its bockstein in an odd-dimensional manifold (4A.39).

4A.1 Definition (see (3.1) and (3.7)). Let D^{k+1} be an oriented (k+1)-disc and choose disjoint oriented discs $D_1^k,\dots,D_t^k \subseteq \partial D^{k+1}$. Attach to each D_i^k a copy of $D_i^k \times [-1,0]$ along $D_i^k \times \{0\}$, where $[-1,0]$ denotes the interval from -1 to 0. Let $\bar{\mathfrak{M}}_t^{k+1}$ be the space obtained by identifying the $D_i^k \times [-1,0]$ by orientation preserving homeomorphisms h_i, where $h_i: D_i^k \times [-1,0] \overset{\approx}{\to} D_{i+1}^k \times [-1,0]$ is $g_i \times id_{[-1,0]}$, and $g_i: D_i^k \approx D_{i+1}^k$, $i = 1,\dots,t-1$. $\bar{\mathfrak{M}}_t^{k+1}$ is called a relative Moore space. The image of (any) $D_i^k \times \{0\}$ under identification is called the bockstein, $b\bar{\mathfrak{M}}_t^{k+1}$, and the image of $D_i^k \times [-1,0]$ is the collar, $c\bar{\mathfrak{M}}_t^{k+1}$. Similarly, the Moore space $\mathfrak{M}_t^k$ is formed by removing the interiors of oriented discs, $\mathring{D}_1^k,\dots,\mathring{D}_t^k$ from an oriented k-sphere S^k, attaching copies of $\partial D_i^k \times [-1,0]$ along $\partial D_i^k \times \{0\}$, then identifying all the $\partial D_i^k \times [-1,0]$ by orientation-preserving homeomorphisms. The image of $\partial D_i^k \times \{0\}$ is the bockstein, $b\mathfrak{M}_t^k$ and $\partial D_i^k \times [-1,0]$, the collar, $c\mathfrak{M}_t^k$. There is an obvious inclusion $i: \mathfrak{M}_t^k \to \bar{\mathfrak{M}}_t^{k+1}$. The boundary of $\bar{\mathfrak{M}}_t^{k+1}$ is $i(\mathfrak{M}_t^k)$ (cf. Remark (4A.2)) and its interior, $\mathring{\bar{\mathfrak{M}}}_t^{k+1}$, is $\bar{\mathfrak{M}}_t^{k+1} - (i(\mathfrak{M}_t^k) \cup c\bar{\mathfrak{M}}_t^{k+1})$. The interior of $\mathfrak{M}_t^k$, $\mathring{\mathfrak{M}}_t^k$, is

$\mathfrak{M}_t^k - c\mathfrak{M}_t^k$.

4A.2 <u>Remark</u>. (a) $\bar{\mathfrak{M}}_t^{k+1}$ and $\mathfrak{M}_t^k$ are $\mathbb{Z}_t$-manifolds in the sense of [MS] and [BRS]. In this context $\bar{\mathfrak{M}}_t^{k+1}$ is a "manifold-with-boundary" and its "boundary" is $\mathfrak{M}_t^k$. (b) If we were working in the category of smooth manifolds, we would not use the collar $c\mathfrak{M}_t^k$. The collar is a p.l. replacement for the tangent direction given by the sheets coming together at the bockstein. We sometimes omit mention of the collar; however, it is implicit and will be used to describe framing conditions. (c) There are definitions of generalized Moore spaces and relative generalized Moore spaces ((3.1) and (3.7)) including the collar.

4A.3 . Let $C(t)$ be the simplicial complex consisting of the cone on t points $x_1,\ldots,x_t$, union a one-simplex $[-1,0]$, where the union is taken at the cone point, 0:

4A.4 <u>Figure</u>.

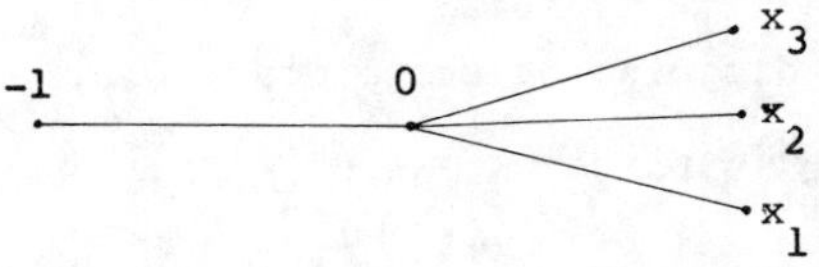

C(3)

Observe that $x \in \bar{\mathfrak{M}}_t^{k+1}$ has a neighborhood p.l. homeomorphic to (a) R^{k+1}, if $x \in \dot{\mathfrak{M}}_t^{k+1}$; (b) R_+^{k+1}, if $x \in \mathfrak{M}_t^k - c\mathfrak{M}_t^k \subseteq \bar{\mathfrak{M}}_t^{k+1}$; (c) $R^k \times C(t)$, if $x \in b\bar{\mathfrak{M}}_t^{k+1} - b\mathfrak{M}_t^k$; or (d) $R_+^k \times C(t)$, if $x \in b\mathfrak{M}_t^k \subseteq \bar{\mathfrak{M}}_t^{k+1}$. Similarly, $x \in \mathfrak{M}_t^k$ has a neighborhood p.l. homeomorphic to (a) R^k, if $x \in \dot{\mathfrak{M}}_t^k$; or (b) $R^{k-1} \times C(t)$, if $x \in b\mathfrak{M}_t^k$.

4A.5 . Let $c\colon C(t) \to R^2$ be the imbedding of $C(t)$ in R^2 given as follows. Referring to Figure (4A.4), identify $[-1,0]$ with the interval $[-1,0]$ on the x-axis in R^2, let x_i be mapped to $z_i = (1,y_i)$, where $y_1 = -t/2$, $y_t = t/2$, and $y_{i+1} - y_i = t/t-1$, $i = 1,\ldots,t-1$, and extend linearly.

4A.6 Figure.

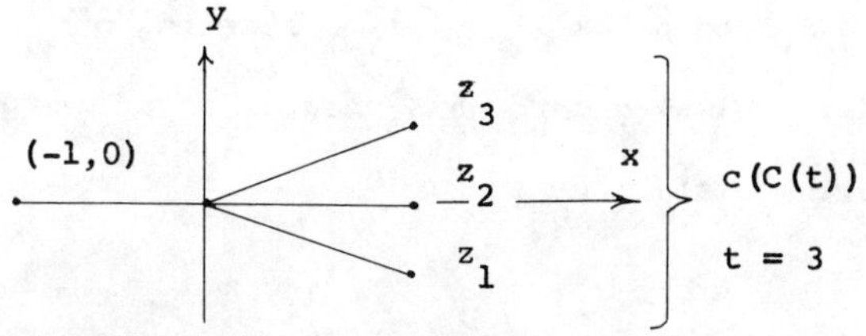

Let $\nu_{C(t)}$ be the regular neighborhood of the imbedding c consisting of the set of all vertical lines of length 1/2 with center at points of $C(t)$.

4A.7 Figure.

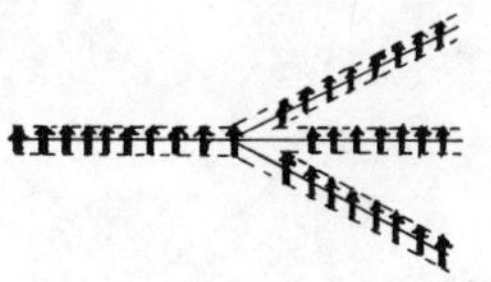

$\nu_{C(t)}$.

The line passing through x, for each $x \in C(t)$, defines a continuous, upward-pointing vector field ε_0 on $C(t)$, transverse to $C(t)$ at x. (At $x = 0$, the field ε_0 is transverse to each line abutting to 0.) We call ε_0 the <u>distinguished field</u> for the imbedding $c\colon C(t) \to R^2$.

4A.8 <u>Proposition</u>. There is a map $F\colon \mathfrak{M}_t^k \to R^{k+1}$, an open set $U \subseteq \mathfrak{M}_t^k$ containing $c\mathfrak{M}_t^k$, and a homeomorphism $k\colon S^{k-1} \times C(t) \overset{\approx}{\to} \bar{U}$ (the closure of U in $\mathfrak{M}_t^k$) such that

(i) $F|\bar{U}$ is an imbedding, $F(c\mathfrak{M}_t^k) \subseteq R_-^{k+1}$, and $F(\bar{U} - (c\mathfrak{M}_t^k - b\mathfrak{M}_t^k)) \subseteq R_+^{k+1}$;

(ii) $(F|c\mathfrak{M}_t^k)\circ(k|S^{k-1} \times [-1,0]) = j \times i\colon S^{k-1} \times [-1,0] \to R^k \times R^1 = R^{k+1}$, where $j\colon S^{k-1} \to R^k$ and $i\colon [-1,0] \to R^1$ are the standard inclusions;

(iii) $(F|\bar{U})\circ k\colon S^{k-1} \times C(t) \to R^{k+1}$ extends to a (codimension-zero) imbedding $S^{k-1} \times \nu_{C(t)} \to R^{k+1}$, thus framing $F|\bar{U} - b\mathfrak{M}_t^k$;

(iv) $F(\bar{U}) \cap F(\mathfrak{M}_t^k - \bar{U}) = \emptyset$ and $F|\mathfrak{M}_t^k - U$ is a framed immersion into R_+^{k+1}, with framing extending that of $F|\bar{U} - b\mathfrak{M}_t^k$.

<u>Proof</u>. Let $S^k \subseteq R^{k+1}$ be the standard imbedding. It is oriented by choosing an outward pointing field η of normal vectors. Let $S_\pm^k = S^k \cap R_\pm^{k+1}$. Choose t-1 disjoint k-discs $D_1,\ldots,D_{t-1}$ near the north

pole in S^k_+ and paths $\ell_1,\ldots,\ell_{t-1}$ starting outward from the center of each disc, passing through S^k_+ near $S^k_+ \cap S^k_-$ and ending at the center of S^k_- (south pole).

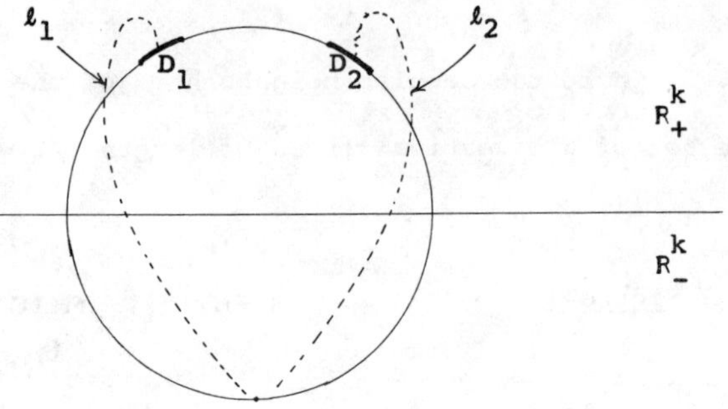

4A.9 Figure. $R^{k-1} = R^k_+ \cap R^k_-$

Construct a regular homotopy of the standard imbedding of S^k in R^{k+1} by dragging each D_i along ℓ_i so that it is identified with S^k_- in orientation-preserving fashion (the normal fields at the identified points have the same direction) and so that the resultant immersion of S^k in R^{k+1} intersects itself in transverse position in R^{k+1}_+.

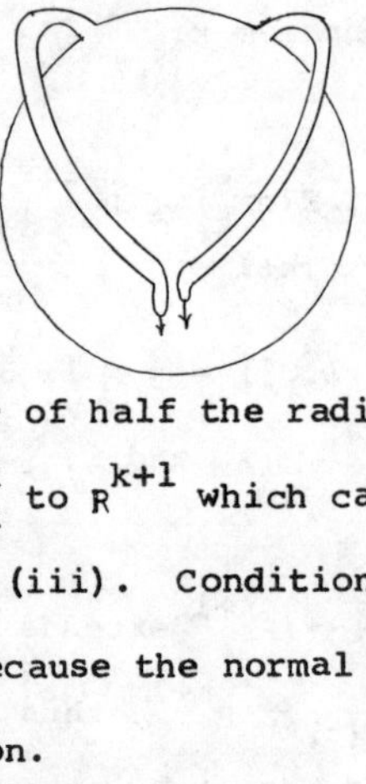

4A.10 Figure.

Cutting out a concentric disc of half the radius from the identified k-disc S^k_- yields a map of $\mathfrak{M}^k_t$ to R^{k+1} which can be arranged to satisfy the local conditions of (ii) and (iii). Condition (i) is clear by construction and (iv) is satisfied because the normal vectors at identified points have the same direction.

Here is the relative version of (4A.8).

4A.11 Proposition. There is a map $\bar{F}\colon \bar{\mathfrak{M}}^{k+1}_t \to R^{k+2}_+$, an open set $V \subseteq \bar{\mathfrak{M}}^{k+1}_t$ containing $c\bar{\mathfrak{M}}^{k+1}_t$ and a homeomorphism $h\colon D^k \times C(t) \xrightarrow{\approx} \bar{V}$ (the closure of V) such that

(i) $\bar{F}$ extends the map constructed in (4A.8) and

$\bar{F}(\mathring{\bar{\mathfrak{M}}}{}_t^{k+1}) \subseteq \overset{\bullet}{R}{}_+^{k+2} = R_+^{k+2} - R^{k+1}$;

(ii) $\bar{F}|\bar{V}$ is an imbedding,

(iii) $(\bar{F}|c\bar{\mathfrak{M}}_t^{k+1})\cdot(h|D^k \times [-1,0]) = \bar{j} \times i\colon D^k \times [-1,0] \to R^{k+1} \times R^1 = R^{k+2}$, where $\bar{j}$ is an imbedding with image $S_+^k = S^k \cap R_+^{k+1}$, and i is the standard inclusion;

(iv) $(\bar{F}|\bar{V})\cdot h\colon D^k \times C(t) \to R^{k+2}$ extends to a (codimension-zero) imbedding $D^k \times \nu_{C(t)} \to R^{k+2}$ thus framing $\bar{F}|(\bar{V} - b\bar{\mathfrak{M}}_t^{k+1})$; and

(v) $\bar{F}(\bar{V}) \cap \bar{F}(\bar{\mathfrak{M}}_t^{k+1} - \bar{V}) = \emptyset$ and $\bar{F}|\bar{\mathfrak{M}}_t^{k+1} - V$ is a framed immersion into R_+^{k+2}, with framing extending that of $\bar{F}|\bar{V} - b\mathfrak{M}_t^{k+1}$.

Proof. Extend the $S^k \subseteq R^{k+1}$ used in the proof of (4A.8) to $S_+^{k+1} = S^{k+1} \cap R_+^{k+2} \subseteq R_+^{k+2}$, and extend the regular homotopy of S^k to one of S_+^{k+1} making the same identifications along $\partial(S_+^{k+1}) = S^k$.

4A.12 _Remark_. The proof of (4A.8) shows that if R^{k+1} as the target of F, then F is homotopic to an imbedding: lift the paths ℓ_i into R^{k+2} to pass over S^k instead of through it. Similarly, $\bar{\mathfrak{M}}_t^{k+1}$ imbeds in R^{k+3}.

4A.13 _Corollary_. There are compact parallelizable manifolds-with-boundary $n\bar{\mathfrak{M}}_t^{k+1}$ and $n\mathfrak{M}_t^k$, of dimension $k+2$ and $k+1$, respectively, together with inclusion j, $\bar{j}$, i, $\bar{i}$, and collapsing maps c and $\bar{c}$ making the following diagram commute

4A.14
$$\begin{array}{ccccc}
\multicolumn{5}{c}{\text{identity}} \\
\mathfrak{M}_t^k & \xrightarrow{\quad j \quad} & n\mathfrak{M}_t^k & \xrightarrow{\quad c \quad} & \mathfrak{M}_t^k \\
\downarrow{\scriptstyle i} & & \downarrow{\scriptstyle \bar{i}} & & \downarrow{\scriptstyle i} \\
\bar{\mathfrak{M}}_t^{k+1} & \xrightarrow{\quad \bar{j} \quad} & n\bar{\mathfrak{M}}_t^{k+1} & \xrightarrow{\quad \bar{c} \quad} & \bar{\mathfrak{M}}_t^{k} \\
\multicolumn{5}{c}{\text{identity}}
\end{array}$$

Here j, c, $\bar{j}$, and $\bar{c}$ are homotopy equivalences.

Proof. The space $n\bar{\mathfrak{M}}_t^{k+1}$ (resp. $n\mathfrak{M}_t^k$) is defined to be the union of

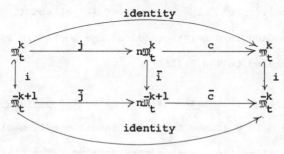

$\nu_{C(t)} \times D^k$ (resp. $\nu_{C(t)} \times S^{k-1}$) with the trivial normal bundle to $\bar{F}|(\bar{\mathfrak{M}}_t^{k+1} - \bar{V})$ (resp. $F|(\mathfrak{M}_t^k - V)$) where $\bar{F}$ and $\bar{V}$ (resp. F and V) are as in Prop. (4A.11) (resp. Prop. (4A.8)). The latter propositions construct codimension zero immersions of $n\mathfrak{M}_t^k$ and $n\bar{\mathfrak{M}}_t^{k+1}$ in Euclidean space, so each has trivial tangent bundle. Using the normal field ϵ_0 constructed in (4A.5), define collapsing maps $\bar{c}'\colon \nu_{C(t)} \times D^k \to C(t) \times D^k$ and $c'\colon \nu_{C(t)} \times S^{k-1} \to C(t) \times S^{k-1}$, which by (4A.11)(v) and (4A.8)(iv) extend to give c and $\bar{c}$ making (4A.14) commute.

4A.15 Definition. Let (N,M) be a p.l. manifold pair. A map $f\colon \mathfrak{M}_t^k \to M$ is called an immersion if it extends to an immersion $F'\colon n\mathfrak{M}_t^k \to M$. A map $\bar{F}\colon (\bar{\mathfrak{M}}_t^{k+1}, \mathfrak{M}_t^k) \to (N,M)$ is an immersion if it extends to an immersion of pairs $\bar{F}'\colon (n\bar{\mathfrak{M}}_t^{k+1}, n\mathfrak{M}_t^k) \to (N,M)$. F or $\bar{F}$ is framed provided F' or $\bar{F}'$ is. A homotopy $H\colon \mathfrak{M}_t^k \times I \to M$ is a regular homotopy if $G = (H,t)\colon \mathfrak{M}_t^k \times I \to M \times I$ extends to an immersion $n\mathfrak{M}_t^k \times I \to M \times I$. A regular homotopy of $\bar{F}$ is defined similarly. The normal bundle to an immersion $F\colon \mathfrak{M}_t^k \to M$ is the regular neighborhood of $\mathfrak{M}_t^k$ in the normal bundle to $F'\colon n\mathfrak{M}_t^k \to M$. The trivial bundle is $n\mathfrak{M}_t^k \times D^\ell$ (compare (4A.16)). This completes the definition.

According to (4A.13), (4A.8), and their proofs, $n\mathfrak{M}_t^k$ is "given" by a vector field ϵ transverse to $\mathfrak{M}_t^k$ in $n\mathfrak{M}_t^k$. This field is induced by the normal field η to $S^k \subseteq R^{k+1}$ in the proof of (4A.8) and (in suitable coordinates) is induced in a neighborhood of the bockstein by the field ϵ_0 of (4A.5), according to (4A.8)(iii).

4A.16 Definition. The field ϵ is called the distinguished field on $\mathfrak{M}_t^k$. If $f\colon \mathfrak{M}_t^k \to M$ is a framed immersion into the p.l. manifold M, the framing of $n\mathfrak{M}_t^k$ is called the complementary framing. The distinguished field and complementary framing define the trivial normal bundle (4A.15). Similar definitions are made for the relative Moore space $\bar{\mathfrak{M}}_t^{k+1}$.

Next we present two useful constructions in $n\mathfrak{M}_t^k$.

4A.17. For $i = 1,\ldots,t$, let $w_i \in [-\frac{1}{8},\frac{1}{8}]$ be such that $w_1 = -\frac{1}{8}$, $w_t = \frac{1}{8}$ and $w_i - w_{i-1} = (t-1)/4$. In the notation of (4A.5) let $L \subseteq \nu_{C(t)} \subseteq R^2$ consist of the union of the t disjoint broken line segments $\ell_i = \{(x,w_i) \in R^2 | x \in [-1,0]\} \cup \{$the line connecting $(0,w_i)$ to $z_i\}$. Each ℓ_i lies in $\nu_{C(t)}$ (see (4A.7)).

4A.18

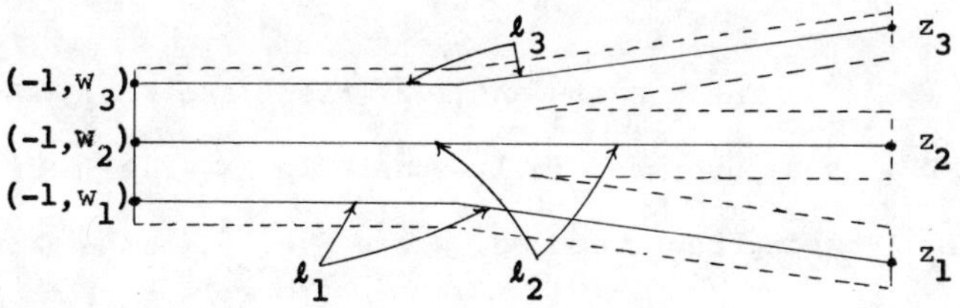

Letting $C(t) \times S^{k-1}$ be identified with a neighborhood of $c\mathfrak{M}_t^k$ in $\mathfrak{M}_t^k$, denote $P = \mathfrak{M}_t^k - (C(t) \times S^{k-1})$, and let $P \cup (L \times S^{k-1})$ denote the union taken along $\cup(z_i \times S^{k-1})$. We call the process of replacing $\mathfrak{M}_t^k$ in $n\mathfrak{M}_t^k$ by $P \cup (L \times S^{k-1})$, <u>pushing $\mathfrak{M}_t^k$ apart at its bockstein</u>. $P \cup (L \times S^{k-1})$ is homeomorphic to S^k - (t k-discs) and has a natural framing induced from that on L in $\nu_{C(t)}$ and the distinguished framing. The opposite process of passing from $P \cup (L \times S^{k-1})$ in $n\mathfrak{M}_t^k$ to $\mathfrak{M}_t^k$ is called "pushing $S^k - (\bigcup_i D_i^k)$ together at its bockstein."

4A.19 <u>The dual Moore space</u>. In the total space $(S^k \times D^m, S^k \times \partial D^m)$ of the trivial disc-and sphere-bundle pair over S^k, the fiber over a point $s_0 \in S^k$ is a pair $(s_0 \times D^m,\ s_0 \times \partial D^m)$ where $s_0 \times D^m$ is transverse or "dual" to the zero section. Analogously, in the trivial bundle $n\mathfrak{M}_t^k \times D^m$ (the normal bundle to the immersion $\mathfrak{M}_t^k \xrightarrow{F} R^{k+1} \hookrightarrow R^{k+m+1}$, where F is constructed in (4A.8)) there is a pair $({}_*\mathfrak{M}_t^{-m+2}, {}_*\mathfrak{M}_t^{m+1}) \subseteq (n\mathfrak{M}_t^k \times D^m,\ \partial(n\mathfrak{M}_t^k \times D^m))$ dual to the canonical inclusion ("zero section"), $\mathfrak{M}_t^k \hookrightarrow n\mathfrak{M}_t^k \times D^m$. (The homological meaning of "dual" is explained by (4B.43)).

These <u>dual Moore spaces</u> are constructed as follows. There is a neighborhood of $b\mathfrak{M}_t^k$ in $\mathfrak{M}_t^k$ homeomorphic to $C(t) \times S^{k-1}$ and ((4A.8)) the regular neighborhood of $C(t) \times S^{k-1}$ in $n\mathfrak{M}_t^k \times D^m$ is $\nu_{C(t)} \times S^{k-1} \times D^m$, which we identify with a subspace of $n\mathfrak{M}_t^k \times D^m$ and include into $R^2 \times R^k \times R^m$ by the imbedding $c\colon C(t) \to R^2$ (4A.5) on the first factor and the standard inclusions on the second and third. Using this imbedding consider the intersection

$$X = (\nu_{C(t)} \times S^{k-1} \times D^m) \cap (R^2 \times s_0 \times R^m)$$

for some $s_0 \in S^{k-1}$. Then X is an $(m+2)$-disc (see Fig. (4A.20)) and $X \cap \partial(n\mathfrak{M}_t^k \times D^m)$ is a punctured $(m+1)$-sphere in ∂X, the $(m+1)$-discs ${}_*B_1^{m+1}, \ldots, {}_*B_t^{m+1}$ having been removed, where $\bigcup_i {}_*B_i^{m+1} = \partial X - (X \cap \partial(n\mathfrak{M}_t^k \times D^m))$.

4A.20 <u>Figure</u>.

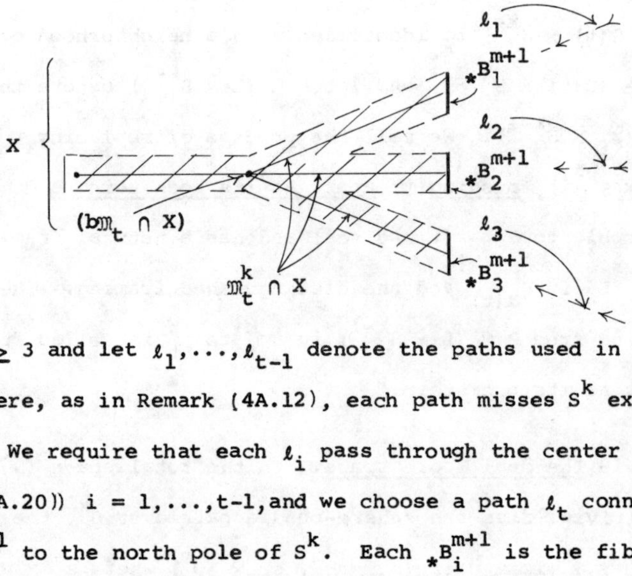

Assume now that $m \geq 3$ and let $\ell_1, \ldots, \ell_{t-1}$ denote the paths used in the proof of (4A.8) where, as in Remark (4A.12), each path misses S^k except at its endpoints. We require that each ℓ_i pass through the center of ${}_*B_i^{m+1}$ (see Fig. (4A.20)) $i = 1, \ldots, t-1$, and we choose a path ℓ_t connecting the center of ${}_*B_t^{m+1}$ to the north pole of S^k. Each ${}_*B_i^{m+1}$ is the fiber over a point of $\mathfrak{M}_t^k$ in the trivial bundle $n\mathfrak{M}_t^k \times D^m$ over $\mathring{\mathfrak{M}}_t^k$. Dragging the ${}_*B_i^{m+1}$ along the "paths in the base," $\ell_i \subset \mathring{\mathfrak{M}}_t^k$, and identifying each ${}_*B_i^{m+1}$ in orientation-preserving fashion with the fiber ${}_*D^{m+1}$ over the north pole, then the union of the track of this dragging process with X yields a framed imbedded relative Moore space ${}_*\bar{\mathfrak{M}}_t^{m+2} \subseteq n\mathfrak{M}_t^k \times D^m$. Its boundary, ${}_*\mathfrak{M}_t^{m+1}$, is $\bar{\mathfrak{M}}_t^{m+2} \cap \partial(n\mathfrak{M}_t^k \times D^m)$, and $b_*\mathfrak{M}_t^{m+2} = {}_*D^{m+1}$, $b_*\mathfrak{M}_t^{m+1} = \partial({}_*D^{m+1})$.

We are now in a position to construct immersions of Moore spaces from surgery data.

4A.21 <u>Theorem</u>. Let $(g,b): (N^n, \nu_N) \to (Y^n, \xi)$ be a normal map and let $\alpha \in \pi_{k+1}(g;\mathbb{Z}_t)$ be represented (3.13) by the commutative square (where $k + 1 < n - 1$):

$$\begin{array}{ccc} \mathfrak{M}_t^k & \xrightarrow{\quad F \quad} & N^n \\ \downarrow i & & \downarrow g \\ \bar{\mathfrak{M}}_t^{k+1} & \xrightarrow{\quad \bar{F} \quad} & Y^n \end{array}$$

Then α determines a framed immersion $F': \mathfrak{M}_t^k \to N$ which is homotopic to F and is unique up to regular homotopy.

<u>Proof</u>. From (4A.14) we have the following commutative diagram:

4A.22
$$\begin{array}{ccccc} \mathfrak{M}_t^k & \xrightarrow{\quad j \quad} & n\mathfrak{M}_t^k & \xrightarrow{\quad F\circ c \quad} & N^n \\ \downarrow i & & \downarrow \bar{i} & & \downarrow g \\ \bar{\mathfrak{M}}_t^{k+1} & \xrightarrow{\quad \bar{j} \quad} & n\bar{\mathfrak{M}}_t^{k+1} & \xrightarrow{\quad \bar{F}\circ\bar{c} \quad} & Y^n \end{array}$$

The bundle map b induces a stable trivialization T of $\tau_{N^n} + g^*\xi$ ([W2]). Pulling this trivialization back by $F\circ c$ we obtain a stable trivialization $(F\circ c)^*T$ of

4A.23
$$(F\circ c)^*\tau_{N^n} + (F\circ c)^*g^*\xi \cong (F\circ c)^*\tau_{N^n} + \bar{i}^*(\bar{F}\circ\bar{c})^*\xi.$$

By (3.9) $\bar{\mathfrak{M}}_t^{k+1}$, and hence $n\bar{\mathfrak{M}}_t^{k+1}$, has the homotopy type of a bouquet of circles; by (3.13) $\bar{F}$ sends a set $K \subseteq \bar{\mathfrak{M}}_t^{k+1}$ to the base point of Y, where K carries $\pi_1\bar{\mathfrak{M}}_t^{k+1}$. The commutativity of (4A.22) thus yields a well defined stable trivialization of $(F\circ c)^*g^*\xi$, and by (4A.23), one of $(F\circ c)^*\tau_{N^n}$ also. Since $\tau_{n\mathfrak{M}_t^k}$ is trivial, there is a stable bundle monomorphism $\tau_{n\mathfrak{M}_t^k \times D^{n-k-1}} \to (F\circ c)^*\tau_{N^n}$.

By immersion theory ([HP],[H]) this defines an immersion of $n\mathfrak{M}_t^k \times D^{n-k-1}$ in N^n whose regular homotopy class depends only on α. This is by definition a framed immersion $F'\colon \mathfrak{M}^k_t \to N^n$, homotopic to F by the commutativity of (4A.22) and the fact that j is a homotopy equivalence.

The following relative version of (4A.21) uses the relative form of immersion theory.

4A.24 Theorem: Let $(g;b)\colon (N^n,\partial N^n;\nu_N) \to (Y,X;\xi)$ be a normal map of pairs and let $\alpha \in \pi_{k+2}(\Phi;\mathbb{Z}_t)$ be given, where Φ is the square of maps and spaces

$$\begin{array}{ccc} \partial N^n & \xrightarrow{\;g|\partial N^n\;} & X \\ \downarrow & & \downarrow \\ N & \xrightarrow{\;g\;} & Y \end{array},$$

$k+2 < n+1$, $\bar{F}\colon (\bar{\mathfrak{M}}^{k+1}_t,\mathfrak{M}^k_t) \to (N^n,\partial N^n)$ represents $\partial\alpha \in \pi_{k+1}(N^n,\partial N^n;\mathbb{Z}_t)$, and $\partial\colon \pi_{k+2}(\Phi;\mathbb{Z}_t) \to \pi_{k+1}(N^n,\partial N^n;\mathbb{Z}_t)$. Then α determines a framed immersion $\bar{F}'\colon (\bar{\mathfrak{M}}^{k+1}_t,\mathfrak{M}^k_t) \to (N^n,\partial N^n)$ in the homotopy class of $\bar{F}$, unique up to regular homotopy.

The following corollary of the above proofs assumes less, but has no uniqueness clause in its conclusion.

4A.25 Corollary. Given a normal map $(g;b)\colon (N^n,\nu_{N^n}) \to (Y,\xi)$ and a commutative diagram

$$\begin{array}{ccc} \mathfrak{M}^k_{\{t_1,\ldots,t_m\}} & \xrightarrow{\;F\;} & N^n \\ {\scriptstyle i}\downarrow & & \downarrow{\scriptstyle g} \\ \bar{\mathfrak{M}}^{k+1}_{\{t_1,\ldots,t_m\}} & \xrightarrow{\;\bar{F}\;} & Y \end{array}$$

where i is the natural inclusion and $k+1 < n-1$, there is a framed immersion $F'\colon \mathfrak{M}^k_{\{t_1,\ldots,t_m\}} \to N^n$ homotopic to F.

This completes our discussion of immersion theory for Moore spaces. We can now study their point intersections.

4A.26. Let $F\colon \mathfrak{M}^k_t \to M^{2k}$ and $G\colon \mathfrak{M}^k_s \to M^{2k}$ be immersions of Moore spaces into the p.l. manifold M^{2k} such that $F(\mathfrak{M}^k_t) \cap G(b\mathfrak{M}^k_s) = \emptyset = F(b\mathfrak{M}^k_t) \cap G(\mathfrak{M}^k_s)$,

$G(\dot{\mathfrak{M}}_s^k)$ intersects itself and $F(\dot{\mathfrak{M}}_t^k)$ transversely in isolated points, and $F(\dot{\mathfrak{M}}_t^k)$ intersects itself transversely in isolated points. (Given immersions F and G, the intersections can always be so arranged up to regular homotopy.) Because F and G induce the trivial homomorphism on π_1, we can define the algebraic chain intersection, denoted

4A.27 $$F(\mathfrak{M}_t^k) \pitchfork G(\mathfrak{M}_s^k) \in \mathbf{Z}\pi$$

and algebraic self-intersection

4A.28 $$F(\mathfrak{M}_t^k) \pitchfork F(\mathfrak{M}_t^k) \in \mathbf{Z}\pi/J_\epsilon$$

where $\epsilon = (-1)^{k+1}$ (see (2.5)(d) for J_ϵ). (Set theoretic intersection continues to be denoted "$\cap$".) This is done exactly as in [W2,§5]. These intersections are not invariants of regular homotopy as the following fundamental proposition shows.

4A.29 Proposition. Suppose that, in the context of (4A.26), $F(\mathfrak{M}_t^k) \pitchfork G(\mathfrak{M}_s^k) \in t(\mathbf{Z}\pi)$. Then G is regularly homotopic to $G': \mathfrak{M}_s^k \to M^{2k}$ where $F(\mathfrak{M}_t^k) \cap G'(\mathfrak{M}_s^k) = \emptyset$. If $F(\mathfrak{M}_t^k) \pitchfork F(\mathfrak{M}_t^k) \in t(\mathbf{Z}\pi/J_\epsilon)$, $\epsilon = (-1)^{k+1}$, then F is regularly homotopic to an imbedding.

Proof. Let $F(\mathfrak{M}_t^k) \pitchfork G(\mathfrak{M}_s^k) = \Sigma\, n_g g \in \mathbf{Z}\pi$. By assumption, $t|n_g$, for all $g \in \pi$. Let P_i be points of intersection contributing the same element $g = g_{P_i}$, and the same ϵ_{P_i}, to the intersection number (where $\epsilon_{P_i} = \pm 1$ is the sign of intersection at P_i), $i = 1,\dots,t$. Choose a point $Q \in F(b\mathfrak{M}_t^k)$ and, for each $i = 1,\dots,t$, a different sheet S_i of $\mathfrak{M}_t^k$ abutting to $b\mathfrak{M}_t^k$ at Q. Let a loop ℓ_n be described as follows. For $i = 1,\dots,t$, take paths h_i from P_i to Q so that, just before reaching Q, h_i is on S_i; for $n = 1,\dots,t-1$ let $\bar{h}_n$ be a path in $G(\dot{\mathfrak{M}}_s^k)$ connecting P_n to P_{n+1} and containing no other intersection points, and let $\ell_n = h_{n+1}^{-1}\bar{h}_n h_n$.

4A.30 Figure.

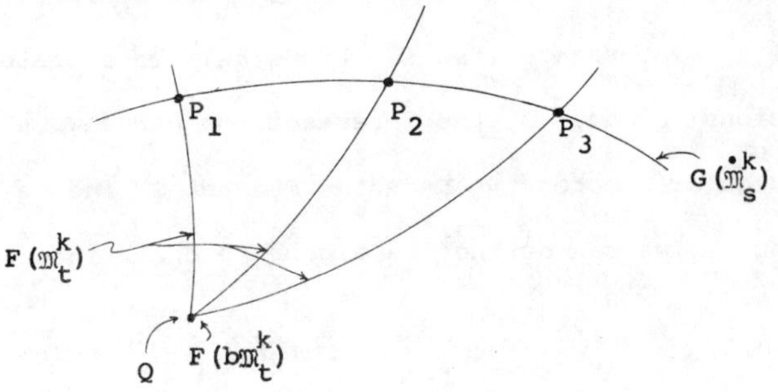

Since $g_{P_i} = g_{P_j}$, each $\ell_n \simeq 1$, $n = 1,\ldots,t-1$. Span an imbedded 2-disc D_n across ℓ_n, so that $\partial D_n = \ell_n$, $\dot{D}_n \cap F(\mathfrak{M}_t^k) = \emptyset = \dot{D}_n \cap G(\mathfrak{M}_s^k)$. For each $n = 1,\ldots,t-1$, push a neighborhood H_n of $\bar{h}_n$ in $G(\dot{\mathfrak{M}}_s^k)$ along D_n until H_n, P_n and P_{n+1} are all in a small Euclidean neighborhood of Q in M^{2k}.

Since F is an immersion, near Q, F can be written (in suitable coordinates) in the form $R^{k-1} \times C(t) \xrightarrow{i\times c} R^{2k-2} \times R^2$, where i is the standard inclusion and c is defined in (4A.5). It has been arranged above that each P_i is in $R^{2k-2} \times R^2$. Arrange next that each P_i lies in a cross section $(i\times c)(\{r\} \times C(t))$, where $r \in R^{k-1}$, $(i\times c)(r\times\{0\}) = Q$ and $\{0\} \in C(t)$, and that each $\bar{h}_i$ also lies in the cross section. Since the signs of intersections ϵ_{P_i} all agree, we may make the regular neighborhood H of $\cup\bar{h}_i$ in $G(\dot{\mathfrak{M}}_s^k)$ transverse to $\{i(r)\} \times R^2$ and so that $H \cap (\{i(r)\} \times R^2) = \bigcup_i \bar{h}_i$.

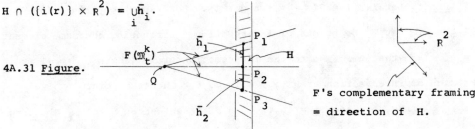

4A.31 Figure.

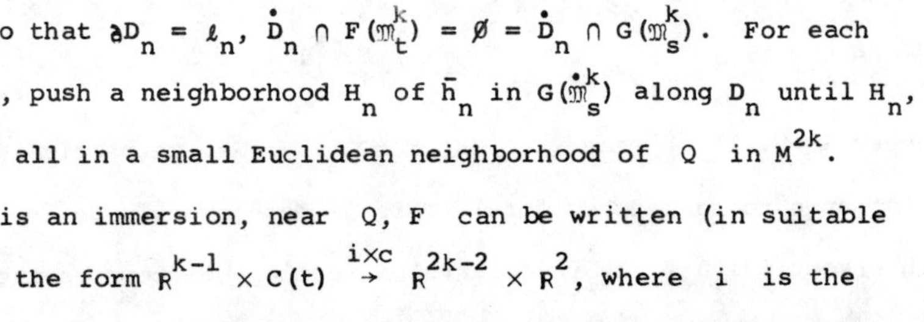

It is now possible to push $G(\mathfrak{M}_s^k)$ off $F(\mathfrak{M}_t^k)$ at Q, moving only a small neighborhood of H. By induction the proof is complete.

In case $F(\mathfrak{M}_t^k) \pitchfork F(\mathfrak{M}_t^k) \in t(\mathbb{Z}\pi/J_\epsilon)$ and $\Sigma n_g g \in \mathbb{Z}\pi$ is a representative, we may assume $t|n_g$, if $g^2 \neq 1$, and for $g^2 = 1$, $t|(n_g + m(1-(-1)^k))$ for some $m \in \mathbb{Z}$. If k is even then $t|n_g$ for all g and we may argue as above to imbed $\mathfrak{M}_t^k$. For k odd, introduce cancelling pairs P, Q of self-intersections ($\epsilon_P = \epsilon_Q$) without changing $\lambda = F(\mathfrak{M}_t^k) \pitchfork F(\mathfrak{M}_t^k) \in \mathbb{Z}\pi/J_\epsilon$. In this way a representative $\Sigma n_g g \in \mathbb{Z}\pi$ for λ can be found for which $t|n_g$, for all $g \in \pi$. Once again the argument above applies.

4A.32 <u>Proposition</u>. Let $F,G: \mathfrak{M}_t^k \to M^{2k}$ be transverse, framed immersions as in (4A.26) ($s = t$). Then there is a framed immersion $\bar{F}: \mathfrak{M}_{st}^k \to M^{2k}$ for which (i) $\bar{F}|b\mathfrak{M}_{st}^k = F|b\mathfrak{M}_t^k$ (each bockstein is identified with a $(k-1)$-sphere), (ii) $\bar{F}(\mathfrak{M}_{st}^k) \pitchfork G(\mathfrak{M}_t^k) \in s(\mathbb{Z}\pi)$, and (iii) $\bar{F}(\mathfrak{M}_{st}^k) \pitchfork \bar{F}(\mathfrak{M}_{st}^k) \in s^2(\mathbb{Z}\pi/J_{(-1)^{k+1}})$.

<u>Proof</u>. Let ϵ denote the distinguished framing (4A.16). At the bockstein F is locally of the form $i \times c: R^{k-1} \times C(t) \to R^{2k-2} \times R^2$ where $i: R^{k-1} \to R^{2k-2}$ is the standard inclusion, $c: C(t) \to R^2$ is given in (4A.5) and ϵ is given by $\nu_{C(t)}$ (4A.7), the upward pointing normals to $c(C(t))$ in R^2. Let the length $|\epsilon|$ of ϵ approach zero near the bockstein to form a new one-frame field $\tilde{\epsilon}$ having length zero on $b\mathfrak{M}_t^k$. Let $\tilde{F}_i: \mathfrak{M}_t^k \to M^{2k}$ be the framed immersion obtained by pushing the image of F along $\tilde{\epsilon}$ a distance $i|\tilde{\epsilon}|/(s-1)$, $i = 1,\ldots,s-1$:

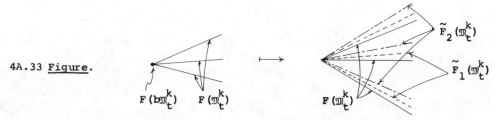

4A.33 <u>Figure</u>.

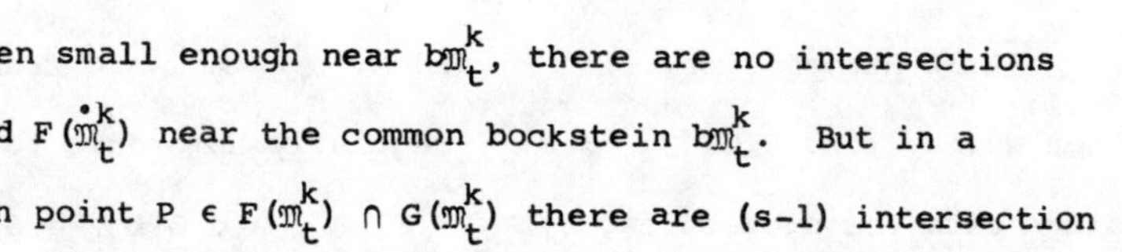

If $|\tilde{\epsilon}|$ is chosen small enough near $b\mathfrak{M}_t^k$, there are no intersections among the $\tilde{F}_i(\dot{\mathfrak{M}}_t^k)$ and $F(\dot{\mathfrak{M}}_t^k)$ near the common bockstein $b\mathfrak{M}_t^k$. But in a neighborhood of each point $P \in F(\mathfrak{M}_t^k) \cap G(\mathfrak{M}_t^k)$ there are $(s-1)$ intersection points $P_i \in \tilde{F}_i(\mathfrak{M}_t^k) \cap G(\mathfrak{M}_t^k)$ such that $g_{P_i} = g_P$ and the sign of intersection at P_i equals that of P, $i = 1,\ldots,s-1$; there are no other intersections of the $\tilde{F}_i(\mathfrak{M}_t^k)$ with $G(\mathfrak{M}_t^k)$.

4A.34 Figure.

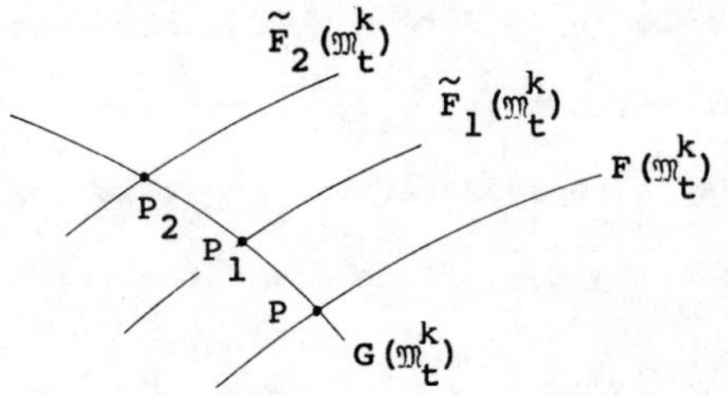

$s = 3$

To construct $\bar{F}: \mathfrak{M}_{st}^k \to M^{2k}$, take interior connected sum of $F(\mathfrak{M}_t^k)$ with $\tilde{F}_1(\mathfrak{M}_t^k)$, and of $\tilde{F}_i(\mathfrak{M}_t^k)$ with $\tilde{F}_{i+1}(\mathfrak{M}_t^k)$, $i = 1,\dots,s-2$ (avoiding points of intersection). This space in M^{2k} is clearly the image of an interior connected sum of punctured k-spheres, where by construction the boundary components are all mapped by orientation-preserving homeomorphisms to $F(b\mathfrak{M}_t^k)$. This is the image of a Moore space of order st, and $\bar{F}$ is thus defined to be $F \mathbin{\hat{\#}} \tilde{F}_1 \mathbin{\hat{\#}} \dots \mathbin{\hat{\#}} \tilde{F}_{t-1}$ where "$\hat{\#}$" means connected sum along the interior and orientation-preserving identification along the bockstein. This yields (i) and (ii) of (4A.32). The proof of (iii) is the same except that the replication process above replaces each double point P of F by s^2 intersection points (compare [W1, p. 232]).

4A.35 Definition-Proposition. The process of replacing F by $\bar{F}$ in the previous proposition is called increasing the order of the bockstein. If $\{\cdot\}$ denotes homotopy class, then in the notation of (3.15), $r_s\{F\} = \{\bar{F}\}$.

Proof. Immediate from the construction of r_s.

4A.36 Proposition. If F and $G: \mathfrak{M}_t^k \to M^{2k}$ are framed immersions, then we may find $\bar{F},\bar{G}: \mathfrak{M}_{t^2}^k \to M^{2k}$, regularly homotopic to disjoint imbeddings with the properties, $F|b\mathfrak{M}_t^k = \bar{F}|b\mathfrak{M}_{t^2}^k$ and $G|b\mathfrak{M}_t^k = \bar{G}|b\mathfrak{M}_{t^2}^k$.

Proof. Immediate from (4A.29) and (4A.32).

4A.37. In [W1, Theorem 5.2] Wall constructs "linking" (resp. "self-linking") numbers in $\mathbb{Q}\pi/\mathbb{Z}\pi$ (resp. $\mathbb{Q}\pi/J_\epsilon$) for certain homology classes of finite order. To fix ideas and notation for use in the sequel we recall this construction using Moore spaces. Let $(g;b): (N^{2k+1};\nu_N) \to (Y;\xi)$ be a normal map such that $K_i(N^{2k+1}) = 0$, $i \neq k$. By Poincaré duality $K_k(N^{2k+1})$ is a (finitely generated) $\mathbb{Z}$-torsion $\mathbb{Z}\pi$-module, annihilated by, say, $t \in \mathbb{Z}$. Given $x,y \in K_k(N^{2k+1})$, use (3.17), (4A.21) and general position to construct framed immersions $F,G: \mathfrak{M}_t^{k+1} \to N^{2k+1}$ such that $F(b\mathfrak{M}_t^{k+1}) \cap G(b\mathfrak{M}_t^{k+1}) = \emptyset$, $F(b\mathfrak{M}_t^{k+1}) \cap G(\mathring{\mathfrak{M}}_t^{k+1})$ and $F(b\mathfrak{M}_t^{k+1}) \cap F(\mathring{\mathfrak{M}}_t^{k+1})$ consist of isolated points, and such that $x = [F(b\mathfrak{M}_t^{k+1})]$ and $y = [G(b\mathfrak{M}_t^{k+1})]$, where $[X]$ denotes the homology class of the cycle X. Let $F': \mathfrak{M}_t^{k+1} \to N^{2k+1}$ denote F pushed a non-zero distance in the direction of the first vector of its complementary framing. Then

$$\varphi(x,y) = \frac{1}{t}(F(b\mathfrak{M}_t^{k+1}) \pitchfork G(\mathfrak{M}_t^{k+1})) \in \mathbb{Q}\pi/\mathbb{Z}\pi$$

4A.38 and

$$\psi(x) = \frac{1}{t}(F'(b\mathfrak{M}_t^{k+1}) \pitchfork F(\mathfrak{M}_t^{k+1})) \in \mathbb{Q}\pi/J_\epsilon$$

are called linking and self-linking numbers, respectively, where "$\pitchfork$" means algebraic chain intersection as in (4A.26) and $\epsilon = (-1)^{k+1}$. According to [W1,Theorem 5.2] and Prop. (2.8), the triple $(K_k(N^{2k+1}),\varphi,\psi)$ is a $(-1)^{k+1}$-form (see Def. (2.9)).

4A.39 <u>Proposition</u>. With the above notation let $x_1,\ldots,x_n \in K_k(N^{2k+1})$ satisfy

$$\varphi(x_i,x_j) = 0 \quad , \quad \psi(x_i) = 0$$

for $i,j = 1,\ldots,n$. Then there are framed immersions $F_i: \mathfrak{M}_t^{k+1} \to N^{2k+1}$ such that $x_i = [F_i(b\mathfrak{M}_t^{k+1})]$ and $F_i(b\mathfrak{M}_t^{k+1}) \cap F_j(\mathring{\mathfrak{M}}_t^{k+1}) = \emptyset$, for $i,j = 1,2,\ldots,n$.

<u>Proof</u>. Construct framed immersions $\tilde{F}_i: \mathfrak{M}_t^{k+1} \to N^{2k+1}$ (satisfying the intersection and general position conditions of the discussion in (4A.37))

such that $[\widetilde{F}_i(b\mathfrak{M}_t^{k+1})] = x_i$, $i = 1,\ldots,n$. For the rest of the proof denote $\mathfrak{M}_t^{k+1} = \mathfrak{M}$ and $N^{2k+1} = N$.

Since $\psi(x_i) = 0$, $\frac{1}{t}(\widetilde{F}_i'(b\mathfrak{M}) \pitchfork \widetilde{F}_i(\mathfrak{M})) = \Sigma\, n_g g \in J_\epsilon$, $\epsilon = (-1)^{k+1}$. According to [W1, p. 251], varying $\widetilde{F}_i$ through a regular homotopy $G_i\colon \mathfrak{M} \times I \to N \times I$ such that $G_i|b\mathfrak{M} \times I\colon b\mathfrak{M} \times I \to N \times I$ has self-intersection equal to $-\Sigma\, n_g g$ yields

$$F_i'(b\mathfrak{M}) \pitchfork F_i(\mathfrak{M}) = 0$$

where $\widetilde{F}_i = G_i|\mathfrak{M} \times \{0\}$, $F_i = G_i|\mathfrak{M} \times \{1\}$. (Once $G_i|b\mathfrak{M} \times I$ is constructed, it is easy to use the definition of immersion (4A.15) to supply the "immersion extension property" needed to construct G_i.) By applying the Whitney procedure, $F_i|\mathring{\mathfrak{M}}$ may be modified by a regular homotopy to yield

$$F_i(b\mathfrak{M}) \cap F_i(\mathring{\mathfrak{M}}) = \emptyset,\ i = 1,\ldots,n.$$

Let $j \geq 2$. By hypothesis $F_j(b\mathfrak{M}) \pitchfork F_1(\mathfrak{M}) = \Sigma\, n_g^j g$, where $t|n_g^j$ for all j and g. Now use the technique of Prop. (4A.29) to push $F_j(b\mathfrak{M})$ along $F_1(\mathfrak{M})$ and across $F_1(b\mathfrak{M})$, so that $F_j(b\mathfrak{M})$ no longer intersects $F_1(\mathfrak{M})$, $j = 2,\ldots,n$. We claim that $F_j(\mathfrak{M}) \pitchfork F_1(b\mathfrak{M}) = 0 \in \mathbb{Z}\pi$. To see this, observe that, in general position, $F_j(\mathfrak{M}) \cap F_1(\mathfrak{M})$ is a union of line segments and circles. Taking the boundary of this 1-chain, we obtain

$$\partial(F_j(\mathfrak{M}) \cap F_1(\mathfrak{M})) = 0.$$

Hence $0 = \partial(F_j(\mathfrak{M}) \cap F_1(\mathfrak{M})) = (\partial F_j(\mathfrak{M})) \pitchfork F_1(\mathfrak{M}) \pm F_j(\mathfrak{M}) \pitchfork (\partial F_1(\mathfrak{M})) =$ $tF_j(b\mathfrak{M}) \pitchfork F_1(\mathfrak{M}) \pm F_j(\mathfrak{M}) \pitchfork (tF_1(b\mathfrak{M})) = \pm tF_j(\mathfrak{M}) \pitchfork F_1(b\mathfrak{M})$, so $F_j(\mathfrak{M}) \pitchfork F_1(b\mathfrak{M}) = 0$ as claimed. Again modify $F_j|\mathring{\mathfrak{M}}$ by the Whitney procedure to obtain $F_1(b\mathfrak{M}) \cap F_j(\mathfrak{M}) = \emptyset$, $j = 1,\ldots,n$, without affecting other point intersections or creating new ones. By induction the proof is complete.

Chapter 4B: Conglomerate Moore Spaces

Let $(g;b)\colon (N;\nu_N) \to (Y;\xi)$ be a highly connected surgery problem where $\dim N = 2k$. Then $K_k(N)$ is $\mathbb{Z}[\pi_1 N]$-free and is represented by a framed immersion $\dot{\cup} S^k \to N$, where "$\dot{\cup}$" denotes disjoint union. Let $(f;b)\colon (M;\nu_M) \to (X;\eta)$ be a highly connected local surgery problem (1.21) where $\dim M = 2k + 1$ and $\pi = \pi_1 M$ is finite. Then by (2.8), $K_k(M) \in \mathfrak{D}_F^1$ (it has a short free resolution over $\mathbb{Z}\pi$). This is homologically analogous to the statement that $K_k(N)$ is free over $\mathbb{Z}\pi$, and we seek a geometric analogue of the representability of $K_k(N)$ by framed, immersed spheres. There is a relative version of this example. If $(G;B)\colon (W;\nu_W) \to (Y\times I;\xi\times I)$ is the bordism of (g,b) corresponding to a set of surgeries on i-spheres in N, then $K_{i+1}(W,N)$ is $\mathbb{Z}\pi$-free with basis represented by core discs of the $(i+1)$-handles. Analogously, if $(F;B)\colon (V;\nu_V) \to (X\times I;\eta\times I)$ is a local bordism of $(f;b)$ having handles of index i and $i + 1$, then it is easy to see that $K_{i+1}(V,M)$ has a short free resolution over $\mathbb{Z}\pi$. Here again we need a geometric model for $K_{i+1}(V,M) \in \mathfrak{D}_F^1$ and its resolution, in some sense analogous to that given for the basis of $K_{i+1}(W,N)$ above.

Moore spaces are not adequate to describe elements $S \in \mathfrak{D}_F^1$, since a short free resolution of S, $(\mathbb{Z}\pi)^n \overset{\mu}{\rightarrowtail} (\mathbb{Z}\pi)^n \twoheadrightarrow S$, requires in general a map μ more complicated than multiplication by some $t \in \mathbb{Z}$ (for any choice of bases in the range or domain of μ). The required geometric object is a conglomerate Moore space, introducted in (4B.1). The two examples above are made precise in (4B.26) and (4B.54), respectively. (The second is formulated for bordisms of manifolds without assuming the presence of a normal map.)

Because conglomerates are built up from Moore spaces by connected sum and because the technical difficulties in the passage from Moore spaces to conglomerates are largely combinatorial, we have isolated in Chapter 4A those properties of conglomerates more easily stated for Moore spaces. Chapter 4B begins with the definition of conglomerates and then builds on the definitions and results of 4A. In addition to the results mentioned above, the main points are that linking numbers detect the geometric intersection of a conglomerate with its bockstein ((4B.33) and (4B.36)) and that there is a description of the linking numbers in the trivial sphere bundle over a conglomerate analogous to that of intersection numbers in the trivial sphere bundle over a sphere (4B.40). In the latter result the "dual conglomerate" (4B.31) is the "fiber" over a point.

In Chapter 4B, if E and F denote free $\mathbb{Z}\pi$-modules, together with given $\mathbb{Z}\pi$-bases $\{e_1,\dots,e_n\}$ and $\{f_1,\dots,f_n\}$, respectively, then, as in Chapter Two, we identify a homomorphism $\mu: E \to F$ with its $(n\times n)$-matrix, also denoted μ. The (i,j)-entry of μ is denoted $\mu_{ij} \in \mathbb{Z}\pi$.

4B.1 <u>The conglomerate $\tilde{\mathcal{C}}^k_\mu$ constructed according to the matrix</u> μ. Let $b\tilde{\mathcal{C}}^k_\mu$ denote the disjoint union of $n|\pi|$ oriented $(k-1)$-spheres $S^{k-1}_{g,i}$, $g \in \pi$, $i = 1,\dots,n$. Let $b\tilde{\mathcal{C}}^k_\mu$ have the right π-action given by permuting the spheres: $S^{k-1}_{g,i}\cdot h = S^{k-1}_{gh,i}$. We will construct $n|\pi|$ generalized Moore spaces whose bocksteins are in $b\tilde{\mathcal{C}}^k_\mu$ and whose union with $b\tilde{\mathcal{C}}^k_\mu$ is $\tilde{\mathcal{C}}^k_\mu$. Let $\mu(e_1) = \Sigma f_j\mu_{j1}$ and let $\mu_{11} = \Sigma m^1_g g$. Let $\mathfrak{M}^k_{\{|m^1_g|\}}$ be the generalized Moore space of order $\{|m^1_g| \mid g \in \pi\}$ (see (3.1) and (4A.1)) and attach it to $\bigcup_g S^{k-1}_{g,1} \subseteq b\tilde{\mathcal{C}}^k_\mu$, so that the component of its bockstein corresponding to $|m^1_h|$ is identified with $S^{k-1}_{h,1}$, for each $h \in \pi$, preserving orientation if $m^1_h > 0$, and reversing it if $m^1_h < 0$. (If $m^1_h = 0$, there is no corresponding component of the bockstein.) In general, attach $\mathfrak{M}^k_{\{|m^i_g|\}}$ to $\bigcup_g S^{k-1}_{g,i}$ for $i = 2,\dots,n$, where $\mu_{i1} = \Sigma m^i_g g$; the attaching must respect orientation as in the case $i = 1$ above. Define

$$\mathfrak{M}(e,1) = \mathfrak{M}^k_{\{|m^1_g|\}} \# \dots \# \mathfrak{M}^k_{\{|m^n_g|\}}$$

where "#" denotes interior connected sum. To form $\mathfrak{M}(g,1)$ where $g \in \pi$, carry out the above construction with $\mu_{i1}g$ in place μ_{i1}, $i = 1,\ldots,n$. The permutation π-action on $b\widetilde{\mathcal{C}}^k_\mu$ extends to $\bigcup_{g\in\pi} \mathfrak{M}(g,1)$ (where the union is taken along $b\widetilde{\mathcal{C}}^k_\mu$) by setting $\mathfrak{M}(g,1)\cdot h = \mathfrak{M}(gh,1)$. Finally using the μ_{ij}, for $j \geq 2$, form $\mathfrak{M}(g,j)$ attached along $b\widetilde{\mathcal{C}}^k_\mu$, setting $\mathfrak{M}(g,j)h = \mathfrak{M}(gh,j)$, $g,h \in \pi$, $j = 1,\ldots,n$. Then we set

4B.2
$$\widetilde{\mathcal{C}}^k_\mu = \Big(\bigcup_{g,j} \mathfrak{M}(g,j)\Big) \cup b\widetilde{\mathcal{C}}^k_\mu$$

where the unions are taken along $b\widetilde{\mathcal{C}}^k_\mu$, and $g \in \pi$, $j = 1,\ldots,n$. $\widetilde{\mathcal{C}}^k_\mu$ is a free π-complex, said to be constructed according to the $(n\times n)$-matrix μ, and called a <u>conglomerate Moore space</u>. We set

4B.3
$$\mathcal{C}^k_\mu = \widetilde{\mathcal{C}}^k_\mu/\pi, \qquad b\mathcal{C}^k_\mu = b\widetilde{\mathcal{C}}^k_\mu/\pi, \qquad b_i\mathcal{C}^k_\mu = \bigcup_g S^{k-1}_{g,i}/\pi.$$

Thinking of $S^{k-1}_{g,i}$ as $f_i\cdot g \in F$ and $\mathfrak{M}(h,j)$ as $e_j\cdot h \in E$, we have geometrically realized the matrix μ (or the map $\mu: E \to F$) in a way that is made precise in Cor. (4B.8).

4B.4 <u>Proposition</u>. Keeping the above notation, let $\widetilde{\mathcal{C}} = \widetilde{\mathcal{C}}^k_\mu$ be constructed according to the matrix μ, where $E \overset{\mu}{\rightarrowtail} F \overset{j}{\twoheadrightarrow} S$ is a short free resolution of the $\mathbb{Z}\pi$-module S. Let $B^k_{g,i} \subseteq \dot{\mathfrak{M}}(g,i)$ be a k-disc ($g \in \pi$, $i = 1,\ldots,n$), and let $\widetilde{B} = \bigcup B^k_{g,i}$ and $\partial\widetilde{B} = \bigcup \partial B^k_{g,i}$. Then ($k \geq 3$)

(i) there is a commutative diagram of $\mathbb{Z}\pi$-modules,

4B.5 <u>Diagram</u>.
$$\begin{array}{ccccc}
H_k(\widetilde{\mathcal{C}}, b\widetilde{\mathcal{C}}) & \overset{\mu''}{\rightarrowtail} & H_{k-1}(b\widetilde{\mathcal{C}}) & \overset{j''}{\twoheadrightarrow} & H_{k-1}(\widetilde{\mathcal{C}}) \\
\cong\downarrow a' & & \cong\downarrow b' & & =\downarrow \\
H_{k-1}(\partial\widetilde{B}) & \overset{\mu'}{\rightarrowtail} & H_{k-1}(\widetilde{\mathcal{C}}-\widetilde{B}) & \overset{j'}{\twoheadrightarrow} & H_{k-1}(\widetilde{\mathcal{C}}) \\
\cong\downarrow a & & \cong\downarrow b & & \cong\downarrow c \\
E & \overset{\mu}{\rightarrowtail} & F & \overset{j}{\twoheadrightarrow} & S
\end{array}$$

with short exact rows and all vertical maps isomorphisms.

(ii) $H_i(\widetilde{\mathcal{C}}) \cong \begin{cases} S, & i = k - 1 \\ 0, & i \neq k - 1, 1, 0. \end{cases}$

(iii) $\widetilde{\mathcal{C}} - \mathring{\widetilde{B}} \simeq b\widetilde{\mathcal{C}} \vee$ (bouquet of circles).

<u>Proof</u>. The map μ'' is the boundary map in the homology sequence of $(\widetilde{\mathcal{C}}, b\widetilde{\mathcal{C}})$. The maps μ', b', j'', and j' are induced by inclusions so the upper right square commutes.

Since $\mathfrak{M}(g,i) - \overset{\bullet}{B}{}^k_{g,i}$ collapses to $b\mathfrak{M}(g,i) \vee$ (circles) (see the proof of (3.2)) it follows that $\widetilde{\mathcal{C}} - \mathring{\widetilde{B}}$ collapses to $b\widetilde{\mathcal{C}} \vee$ (circles), and that b' and the natural map $H_k(\widetilde{\mathcal{C}}, b\widetilde{\mathcal{C}}) \to H_k(\widetilde{\mathcal{C}}, \widetilde{\mathcal{C}} - \mathring{\widetilde{B}})$ are isomorphisms. Defining a' to be the composition

$$H_k(\widetilde{\mathcal{C}}, b\widetilde{\mathcal{C}}) \xrightarrow{\cong} H_k(\widetilde{\mathcal{C}}, \widetilde{\mathcal{C}} - \mathring{\widetilde{B}}) \underset{\text{excision}}{\xleftarrow{\cong}} H_k(\widetilde{B}, \partial\widetilde{B}) \xrightarrow{\partial} H_{k-1}(\partial\widetilde{B}),$$

it follows that a' is an isomorphism and that the upper left square commutes.

From the Mayer-Vietoris sequence

4B.6 $$\cdots \to H_{i+1}(\widetilde{\mathcal{C}}) \longrightarrow H_i(\partial\widetilde{B}) \longrightarrow H_i(\widetilde{B}) + H_i(\widetilde{\mathcal{C}} - \widetilde{B}) \longrightarrow H_i(\widetilde{\mathcal{C}}) \longrightarrow \cdots$$

and the isomorphism b', we extract the exact sequence

4B.7 $$H_k(\widetilde{\mathcal{C}}) \rightarrowtail H_{k-1}(\partial\widetilde{B}) \xrightarrow{\mu'} H_{k-1}(\widetilde{\mathcal{C}} - \mathring{\widetilde{B}}) \xrightarrow{j'} H_{k-1}(\widetilde{\mathcal{C}}).$$

$H_{k-1}(\partial\widetilde{B})$ is $\mathbb{Z}\pi$-free with basis $[\partial B_{e,1}], \ldots, [\partial B_{e,n}]$, and (using b') $H_{k-1}(\widetilde{\mathcal{C}} - \mathring{\widetilde{B}})$ is $\mathbb{Z}\pi$-free with basis $[S^{k-1}_{e,1}], \ldots, [S^{k-1}_{e,n}]$ $(b\widetilde{\mathcal{C}} = \cup S^{k-1}_{g,i})$. Since by construction (see the proof of (3.2)) $\mu'[\partial B_{e,i}] = \sum_j [S^{k-1}_{e,j}]\mu_{ji}$, it follows that μ' is injective (since μ is) and that the bottom left square commutes if we set $a[\partial B_{e,i}] = e_i$, $b[S^{k-1}_{e,i}] = f_i$. From this follow the remaining properties in part (i) and the description of $H_*(\widetilde{\mathcal{C}})$ in (ii).

The following is a consequence of the above proof.

4B.8 <u>Corollary</u>. With the assumptions and notation of (4B.4),

$H_k(\tilde{\mathcal{C}}, b\tilde{\mathcal{C}})$ is $\mathbb{Z}\pi$-free with basis $[\mathfrak{M}(e,1)],\ldots,[\mathfrak{M}(e,n)]$ and $\mu''[\mathfrak{M}(e,i)] = \sum_j [S^{k-1}_{e,j}]\mu_{ji}$, where $\mu''\colon H_k(\tilde{\mathcal{C}}, b\tilde{\mathcal{C}}) \to H_{k-1}(b\tilde{\mathcal{C}})$.

4B.9 Definition. $E \xrightarrow{\mu} F \xrightarrow{j} H_{k-1}(\tilde{\mathcal{C}}^k_\mu)$ is the resolution corresponding to $\tilde{\mathcal{C}}^k_\mu$ (or just to $\mathcal{C}^k_\mu$) if $E = H_k(\tilde{\mathcal{C}}^k_\mu, b\tilde{\mathcal{C}}^k_\mu)$ with basis $\{[\mathfrak{M}(e,i)]\}$, $F = H_{k-1}(b\tilde{\mathcal{C}}^k_\mu)$ with basis $\{[S^{k-1}_{e,i}]\}$, $\mu = \mu''$ (see (4B.5) and j is induced by $b\tilde{\mathcal{C}}^k_\mu \to \tilde{\mathcal{C}}^k_\mu$.

4B.10 Remark. $M_n(\mathbb{Z}\pi, \mathbb{Q}\pi)^\times$ (2.42) denotes the set of $(n\times n)$-matrices which have entries in $\mathbb{Z}\pi$ and are $\mathbb{Q}\pi$-invertible. Then the preceeding proposition and corollary show there is a one-to-one correspondence between elements of $M_n(\mathbb{Z}\pi, \mathbb{Q}\pi)^\times$ and conglomerates $\tilde{\mathcal{C}}^k_\mu$ such that $H_{k-1}(\tilde{\mathcal{C}}^k_\mu)$ is torsion. For such conglomerates, we may replace (4B.2) with

4B.11
$$\tilde{\mathcal{C}}^k_\mu = \bigcup_{g,i} \mathfrak{M}(g,i)$$

since, if some $S^{k-1}_{h,j} \not\subset \bigcup_{g,i} \mathfrak{M}(g,i)$, then $[S^{k-1}_{h,j}] \in H_{k-1}(\tilde{\mathcal{C}}^k_\mu)$ is an element of infinite order. More generally, the second sentence of this Remark remains true if the qualifiers concerning $\mathbb{Q}\pi$-invertibility and finiteness of $H_{k-1}(\tilde{\mathcal{C}}^k_\mu)$ are omitted.

4B.12. Given a generalized moore space $\mathfrak{M}^k_{\{t_1,\ldots,t_n\}}$, an (associated) identified Moore space is obtained by identifying the components of the bockstein to $(k-1)$-spheres by homeomorphisms which are not necessarily orientation preserving. Setting

4B.13
$$\mathfrak{M}(i) = \Big(\bigcup_g \mathfrak{M}(g,i)\Big)/\pi,$$

then $\mathfrak{M}(i)$ is an identified Moore space and

4B.14
$$\bigcup_i \mathfrak{M}(i) = \mathcal{C}^k_\mu.$$

4B.15. Let M^m be a p.l. manifold with $\pi_1 M = \pi$. By a map $F\colon \mathcal{C}^k_\mu \to M^m$ we always mean one which is covered by a π-equivariant map $\tilde{F}\colon \tilde{\mathcal{C}}^k_\mu \to \tilde{M}^m$

(the universal cover of M^m):

4B.16
$$\begin{array}{ccc} \widetilde{\mathcal{C}}^k_\mu & \xrightarrow{\widetilde{F}} & \widetilde{M}^m \\ \downarrow q & & \downarrow p \\ \widetilde{\mathcal{C}}^k_\mu/\pi = \mathcal{C}^k_\mu & \xrightarrow{F} & M^m \end{array}$$

In particular, $(F\cdot q)_\#: \pi_1\widetilde{\mathcal{C}}^k_\mu \to \pi_1 M^m$ is trivial.

4B.17. Following the usual convention ([W2]), for any subset $K \subseteq M^m$, $H_*(K)$ means the $\mathbb{Z}\pi$-module $H_*(p^{-1}K)$. Hence if F is an imbedding, $H_*(F\mathcal{C}^k_\mu) \equiv H_*(\widetilde{\mathcal{C}}^k_\mu)$ as $\mathbb{Z}\pi$-modules. (This is sometimes called homology with "local coefficients.") Let $\mathcal{C}^k_\mu \equiv F\mathcal{C}^k_\mu \subset M^m$ be the image of an imbedding F. For each $g \in \pi$ and $i = 1,\ldots,n$, choose a path $\ell_{g,i}$ from the base point $m_0 \in M^m$ to a reference point in $\mathfrak{M}(i)$ such that $\ell_{g,i}$ defines the lifting $\mathfrak{M}(g,i) \hookrightarrow \widetilde{M}^m$ of $\mathfrak{M}(i) \hookrightarrow M^m$. This done, the k-chain $\mathfrak{M}(i)\cdot a$ in M^m is sometimes used for convenience (see [W1], §5) in place of the (equivalent) chain in $\widetilde{M}^m$, $\mathfrak{M}(e,i)\cdot a$. Intersection numbers of the interiors of such chains and of the interiors with manifolds will be counted in $\mathbb{Z}\pi$ in the usual way (see (4A.27) and (4A.28)). Some caution is necessary because $\mathfrak{M}(i)$ may carry elements of $\pi_1 M$: it is sufficient to ensure that paths used in the counting of intersections do not pass through the bockstein.

4B.18. Since $\mathcal{C}^k_\mu = \cup\, \mathfrak{M}(i)$, $\mathfrak{M}(i) = (\bigcup_g \mathfrak{M}(g,i))/\pi$ and $\mathfrak{M}(g,i)$ is by construction the interior connected sum of ordinary Moore spaces, the reader can easily formulate the definitions of "distinguished field" and "complmentary framing" (see (4A.16)) on $\mathcal{C}^k_\mu$. It is only necessary to guarantee that the identifications along the bockstein converting $\mathfrak{M}(g,i)$ to $\mathfrak{M}(i)$ respect the distinguished field and complementary framing. Recall from (4A.13) and (4A.15) that the definition of an immersion of a Moore space was given in terms of a canonical regular neighborhood $n\mathfrak{M}^k_t$, which in turn is given by the distinguished framing (cf. discussion preceding

(4A.16)). From this the reader can supply the definition of "$n\tilde{\mathcal{C}}^k_\mu$", "$n\mathcal{C}^k_\mu$" and of "an immersion $F: \mathcal{C}^k_\mu \to M^m$".

Let $n\tilde{\mathcal{C}}^k_\mu \times D^m$ be the trivial bundle over $\tilde{\mathcal{C}}^k_\mu$. Since π essentially permutes the terms in $\bigcup_{g,i} \mathfrak{M}(g,i) = \tilde{\mathcal{C}}^k_\mu$, there is a free action of π on $n\tilde{\mathcal{C}}^k_\mu \times D^m$ given by correspondingly permuting the distinguished field and fixing the D^m-factor (i.e., this is the trivial action on the "fibers" of $n\tilde{\mathcal{C}}^k_\mu \times D^m$). We then have distinguished and complementary fields on the quotient and

4B.19 $$(n\tilde{\mathcal{C}}^k_\mu \times D^m)/\pi = n\mathcal{C}^k_\mu \times D^m.$$

4B.20. Let $\mu \in M_n(\mathbb{Z}\pi,\mathbb{Q}\pi)^\times$ (see (4B.10)). Following the construction (4B.1) of $\tilde{\mathcal{C}}^k_\mu$ we construct $\tilde{\bar{\mathcal{C}}}^{k+1}_\mu$. Let $b\tilde{\bar{\mathcal{C}}}^{k+1}_\mu = \bigcup_{g,i} D^k_{g,i}$ and follow the recipe of (4B.1) with $\bar{\mathfrak{M}}^{k+1}_{\{|m^i_g|\}}$ in place of $\mathfrak{M}^k_{\{|m^i_g|\}}$, $i = 1,\ldots,n$. This constructs spaces $\bar{\mathfrak{M}}(g,i)$ whose union is

4B.21 $$\tilde{\bar{\mathcal{C}}}^{k+1}_\mu = \bigcup_{g,i} \bar{\mathfrak{M}}(g,i).$$

The inclusions $\mathfrak{M}(g,i) \hookrightarrow \bar{\mathfrak{M}}(g,i)$ define the inclusion $\tilde{\mathcal{C}}^k_\mu \hookrightarrow \tilde{\bar{\mathcal{C}}}^{k+1}_\mu$ and, setting

4B.22 $$\bar{\mathcal{C}}^{k+1}_\mu = \tilde{\bar{\mathcal{C}}}^{k+1}_\mu/\pi,$$

the inclusion $i: \mathcal{C}^k_\mu \hookrightarrow \bar{\mathcal{C}}^{k+1}_\mu$. Once again we may define "immersion $\bar{F}: \bar{\mathcal{C}}^{k+1}_\mu \to M^m$" by analogy with the absolute case above. From the proof of (4A.21) we obtain:

4B.23 Proposition. Suppose given a normal map $(g;b): (N^n;\nu_N) \to (Y;\xi)$ and a commutative diagram (where $k + 1 < n - 1$)

4B.24 $$\begin{array}{ccc} \mathcal{C}^k_\mu & \xrightarrow{F} & N^n \\ {\scriptstyle i}\downarrow & & \downarrow{\scriptstyle g} \\ \bar{\mathcal{C}}^{k+1}_\mu & \xrightarrow{\bar{F}} & Y \end{array}$$

Then there exists an immersion $F': \mathcal{C}^k_\mu \to N^n$ homotopic to F. If a homotopy class of diagrams of the form (4B.24) is specified, F' is unique up to regular homotopy.

Here are the important homological properties of $\bar{\mathcal{C}}_\mu^{k+1}$ relative to $\mathcal{C}_\mu^k$.

4B.25 <u>Proposition</u>. Notation being as above,

(i) $\overset{\approx}{\mathcal{C}}{}_\mu^{k+1}$ has the homotopy type of a bouquet of circles.

(ii) $H_i(\overset{\approx}{\mathcal{C}}{}_\mu^{k+1}, \tilde{\mathcal{C}}_\mu^k) \cong \begin{cases} H_{k-1}(\tilde{\mathcal{C}}_\mu^k), & i = k \\ 0 & i \neq k \end{cases}$

(iii) $\pi_0\tilde{\mathcal{C}}_\mu^k \overset{\cong}{\to} \pi_0\overset{\approx}{\mathcal{C}}{}_\mu^{k+1}$ and $\pi_1\tilde{\mathcal{C}}_\mu^k \overset{\cong}{\twoheadrightarrow} \pi_1\overset{\approx}{\mathcal{C}}{}_\mu^{k+1}$.

<u>Proof</u>. $\overset{\approx}{\mathcal{C}}{}_\mu^{k+1}$ is a union along discs (the $D_{g,i}^k$) of spaces $\bar{\mathfrak{M}}(g,i)$, each homotopy equivalent to a bouquet of circles (3.9). This yields (i), while (ii) and (iii) follow from (i) and the construction of $\overset{\approx}{\mathcal{C}}{}_\mu^{k+1}$.

The following result explains how relative homology is realized by an immersed relative conglomerate.

4B.26 <u>Proposition</u>. Let $(g;b): (N, \partial N; \nu_N) \to (Y, X; \xi)$ be a normal map of pairs where for some $k \geq 3$, $K_i(N, \partial N) = 0 = K_i(\partial N)$, $i < k$, and where g and the natural inclusions induce isomorphisms $\pi_1(\partial N) \cong \pi_1(N) \cong \pi_1(Y) \cong \pi_1(X)$. Let S be a submodule of $K_k(N)$ where $S \in \mathfrak{I}_F^1$. Then there is an $(n\times n)$-matrix μ and a map of squares

$$\text{4B.27} \qquad \left\{\begin{array}{ccc} \mathcal{C}_\mu^k & \hookrightarrow & \bar{\mathcal{C}}_\mu^{k+1} \\ \downarrow & & \downarrow \\ \bar{\mathcal{C}}_\mu^{k+1} & \hookrightarrow & \bar{\mathcal{C}}_\mu^{k+2} \end{array}\right\} \xrightarrow{\underline{\underline{F}} = \begin{pmatrix} F & H \\ \bar{F} & \bar{H} \end{pmatrix}} \Gamma = \left\{\begin{array}{ccc} \partial N & \xrightarrow{g|\partial N} & X \\ \downarrow & & \downarrow \\ N & \xrightarrow{g} & Y \end{array}\right\}$$

where the square on the left corresponds to (3.19) using $\mathcal{C}_\mu^k = \cup\mathfrak{M}^k(i)$, $\bar{\mathcal{C}}_\mu^{k+1} = \cup\bar{\mathfrak{M}}^{k+1}(i)$, $\bar{\mathcal{C}}_\mu^{k+2} = \cup\,\bar{\mathfrak{M}}^{k+2}(i)$; $\bar{F}: (\bar{\mathcal{C}}_\mu^{k+1}, \mathcal{C}_\mu^k) \to (N, \partial N)$, $\bar{H}: (\bar{\mathcal{C}}_\mu^{k+2}, \mathcal{C}_\mu^{k+1}) \to (Y, X)$, and

a) $\operatorname{cok}(\mu) \cong S$ (where μ is viewed as a homomorphism of free modules),

b) $\bar{F}: (\bar{\mathcal{C}}_\mu^{k+1}, \mathcal{C}_\mu^k) \to (N, \partial N)$ is a framed immersion,

c) $\bar{F}_*: H_k(\bar{\mathcal{C}}_\mu^{k+1}, \mathcal{C}_\mu^k) \to H_k(N, \partial N)$ is an injection whose image is S.

<u>Proof</u>. It suffices by (4B.8), (4B.25)(ii) and the relative version of (4B.23) to find μ satisfying a) and construct the map (4B.27) to satisfy c). By the Hurewicz theorem (as generalized by Namioka [Na]) there are elements

$$\alpha_i \in \pi_{k+1}(\Gamma) = \pi_{k+1}(\widetilde{\Gamma}), \quad i = 1,\dots,n,$$

whose images under the Hurewicz map generate $S \subseteq K_k(N,\partial N) = H_{k+1}(\widetilde{\Gamma})$ as a $\mathbb{Z}\pi$-module, where $\widetilde{\Gamma}$ denotes the square of universal covers of Γ. (See [Hi] for the definition of the homology of a square of spaces.)

Working in $\widetilde{\Gamma}$, let

4B.28
$$a_{i,g}: \left\{ \begin{array}{ccc} S^{k-1}_{i,g} & \rightarrow & D^{k}_{i,g} \\ \downarrow & & \downarrow \\ D^{k}_{i,g} & \rightarrow & D^{k+1}_{i,g} \end{array} \right\} \rightarrow \widetilde{\Gamma}$$

be a map representing $\alpha_i \cdot g \in \pi_{k+1}(\widetilde{\Gamma})$, $i = 1,\dots,n$, $g \in \pi$. Hence if $\dot{\cup}$ denotes disjoint union and $\widetilde{S}_{i,g}$ denotes the domain square of $a_{i,g}$,

$$H_{k+1}(\dot{\bigcup_{i,g}} \widetilde{S}_{i,g}) \xrightarrow{\dot{\cup} a_{i,g}} S \subseteq H_{k+1}(\widetilde{\Gamma})$$

is a surjection whose kernel F is $\mathbb{Z}\pi$-free by Shanuel's Lemma [Ma p.101], since $S \in \mathfrak{P}^1_F$ and $H_{k+1}(\dot{\bigcup_{i,g}} \widetilde{S}_{i,g})$ is $\mathbb{Z}\pi$-free. Let $\{f_1,\dots,f_n\}$ be a $\mathbb{Z}\pi$-basis for F and let μ be the $(n\times n)$-matrix of the map $F \hookrightarrow H_{k+1}(\dot{\bigcup_{i,g}} \widetilde{S}_{i,g})$ with respect to $\{f_1,\dots,f_n\}$ and the $\mathbb{Z}\pi$-basis $\{[\widetilde{S}_{1,e}],\dots,[\widetilde{S}_{n,e}]\}$ for $H_{k+1}(\dot{\bigcup_{i,g}} \widetilde{S}_{i,g})$. For each fixed j we have

4B.29
$$f_j = \sum_{g,i} t^g_{ij}[\widetilde{S}_{i,g}]$$

where $t^g_{ij} \in \mathbb{Z}$ and $\mu_{ij} = \sum_g t^g_{ij} g$. By a straightforward generalization of (3.10) there is a map

4B.30
$$\underline{\underline{F}}(e,j) = \left\{ \begin{array}{ccc} F(e,j) & \rightarrow & H(e,j) \\ \downarrow & & \downarrow \\ \bar{F}(e,j) & \rightarrow & \bar{H}(e,j) \end{array} \right\} : \left\{ \begin{array}{ccc} \mathfrak{M}^k(e,j) & \rightarrow & \bar{\mathfrak{M}}^{k+1}(e,j) \\ \downarrow & & \downarrow \\ \bar{\mathfrak{M}}^{k+1}(e,j) & \rightarrow & \bar{\mathfrak{M}}^{k+2}(e,j) \end{array} \right\} \rightarrow \widetilde{\Gamma}$$

($e \in \pi$ the identity) which extends the disjoint union of the $a_{i,g}$'s ($\underline{\underline{F}}(e,j)$ restricted to the square of bocksteins) in (4B.28). (Compare the statement of (3.10).) Carrying out the same procedure with $f_j \cdot g$ construct $\underline{\underline{F}}(g,j)$: $\left(\begin{array}{ccc} \mathfrak{M}^k(g,j) & \hookrightarrow & \bar{\mathfrak{M}}^{k+1}(g,j) \\ \downarrow & & \downarrow \\ \bar{\mathfrak{M}}^{k+1}(g,j) & \hookrightarrow & \bar{\mathfrak{M}}^{k+2}(g,j) \end{array}\right) \to \tilde{\Gamma}$ requiring $\bar{F}(g,j) = \bar{F}(e,j)g$ and $\bar{H}(g,j) = \bar{H}(e,j)g$. Having done this for each j, by construction the union along appropriate bockstein components

$$\bigcup_{g,j} \left\{ \begin{array}{ccc} \mathfrak{M}^k(g,j) & \hookrightarrow & \bar{\mathfrak{M}}^{k+1}(g,j) \\ \downarrow & & \downarrow \\ \bar{\mathfrak{M}}^{k+1}(g,j) & \hookrightarrow & \bar{\mathfrak{M}}^{k+2}(g,j) \end{array} \right\} \text{ is } \left\{ \begin{array}{ccc} \tilde{\mathcal{C}}^k_\mu & \hookrightarrow & \tilde{\bar{\mathcal{C}}}^{k+1}_\mu \\ \downarrow & & \downarrow \\ \tilde{\bar{\mathcal{C}}}^{k+1}_\mu & \hookrightarrow & \tilde{\bar{\mathcal{C}}}^{k+2}_\mu \end{array} \right\}$$

Setting $\tilde{\underline{\underline{F}}} = \bigcup_{g,j} \underline{\underline{F}}(g,j)$ furnishes a map whose quotient (by the π-action) is easily verified to satisfy c).

4B.31 The dual conglomerate. Let μ be an $(n\times n)$-matrix and $\bar{\mu}$ its conjugate transpose (as in (2.2)). We construct a pair of conglomerates $\bar{\mathcal{C}}^{m+2}_{\bar{\mu}}$ and $\mathcal{C}^{m+1}_{\bar{\mu}}$ such that $\delta: (\bar{\mathcal{C}}^{m+2}_{\bar{\mu}}, \mathcal{C}^{m+1}_{\bar{\mu}}) \hookrightarrow (n\mathcal{C}^k_\mu \times D^m, \partial(n\mathcal{C}^k_\mu \times D^m))$, where $m \geq 3$. If $n = 1$, and $\mu = te$ where $t \in \mathbb{Z}^+$ and $e \in \pi$ is the identity, then $(\bar{\mathcal{C}}^{m+2}_{\bar{\mu}}, \mathcal{C}^{m+1}_{\bar{\mu}}) = ({}_*\bar{\mathfrak{M}}^{m+2}_t, {}_*\mathfrak{M}^{m+1}_t)$ constructed in (4A.19).

In the notation of (4B.1), let $\tilde{\mathcal{C}}^k_\mu = \bigcup_{g,i} \mathfrak{M}(g,i)$ and $b\tilde{\mathcal{C}}^k_\mu = \bigcup_{g,i} S^{k-1}_{g,i}$, where each $\mathfrak{M}(g,i)$ is a generalized Moore space. Let $m(e,i) \in \mathring{\mathfrak{M}}(e,i)$. The fiber over $m(e,i)$ in the trivial bundle $n\mathfrak{M}(e,i) \times D^m$ (see (4A.15) and (4A.16)) is an $(m+1)$-disc, ${}_*D^{m+1}_{e,i}$. The fiber over any other point in $\mathring{\mathfrak{M}}(e,i)$ is isotopic in a fiber-preserving way to ${}_*D^{m+1}_{e,i}$, so the latter is essentially unique. Let $m(g,i) = m(e,i)\cdot g \in \mathring{\mathfrak{M}}(g,i)$ and let ${}_*D^{m+1}_{g,i} = {}_*D^{m+1}_{e,i}\cdot g \in n\mathfrak{M}(g,i) \times D^m$, for all $g \in \pi$ and $i = 1,\ldots,n$.

Fix an integer i and, following the procedure and notation of (4A.19), take the intersection X of an approppiate $(m+2)$-plane with a local model of a nieghborhood of the bockstein component $S^{k-1}_{e,i}$ (see Figure (4A.20)). Each $(m+1)$-disc ${}_*B^{m+1}_r$ in ∂X is the fiber in $n\mathfrak{M}(h,j) \times D^m$ over a point in $\mathring{\mathfrak{M}}(h,j)$ for some $h \in \pi$ and some $j = 1,\ldots,n$. If we drag each ${}_*B^{m+1}_r$ to ${}_*D^{m+1}_{h,j}$ through fibers along a path ℓ_r in the base $\mathfrak{M}(h,j)$, then the union of the track with X will be an $(m+2)$-dimensional relative generalized Moore space ${}_*\bar{\mathfrak{M}}(e,i) \subseteq n\tilde{\mathcal{C}}^k_\mu \times D^m$, whose bockstein $b_*\bar{\mathfrak{M}}(e,i)$ is a subset of the disjoint union $\bigcup_{h,j} {}_*D^{m+1}_{h,j}$. More precisely, $b_*\bar{\mathfrak{M}}(e,i)$ is the disjoint union of those ${}_*D^{m+1}_{h,j}$ such that $\mathfrak{M}(h,j)$ abuts to $S^{k-1}_{e,i}$. Thus, ${}_*\bar{\mathfrak{M}}(e,i)$ abuts to ${}_*D^{m+1}_{h,j}$ with the same multiplicity as that with which $\mathfrak{M}(h,j)$ abuts to $S^{k-1}_{e,i}$. By construction, the latter is the coefficient of $e \in \pi$ in $\mu_{ji}h$, which equals the coefficient of h in $\overline{\mu_{ji}}$, the (i,j)-entry in $\bar{\mu}$. Setting ${}_*\bar{\mathfrak{M}}(g,i) = {}_*\bar{\mathfrak{M}}(e,i)\cdot g$ for each $g \in \pi$, we see that $\bigcup_{g,i} {}_*\bar{\mathfrak{M}}(g,i) = \tilde{\mathcal{C}}^{m+2}_{\bar{\mu}}$, the conglomerate constructed according to $\bar{\mu}$. Define $\tilde{\mathcal{C}}^{m+1}_{\bar{\mu}}$ to be $\tilde{\mathcal{C}}^{m+2}_{\bar{\mu}} \cap \partial(n\mathcal{C}^k_\mu \times D^m)$, $\mathcal{C}^{m+2}_{\bar{\mu}} = \tilde{\mathcal{C}}^{m+2}_{\bar{\mu}}/\pi$ and $\mathcal{C}^{m+1}_{\bar{\mu}} = \tilde{\mathcal{C}}^{m+1}_{\bar{\mu}}/\pi$. From (2.15) and the fact that μ is the conjugate transpose of μ we have

4B.32
$$H_m(\tilde{\mathcal{C}}^{m+1}_{\bar{\mu}}) \cong H_{k-1}(\tilde{\mathcal{C}}^k_\mu)^\wedge,$$

where $M^\wedge = \mathrm{Hom}_{\mathbb{Z}\pi}(M, \mathbb{Q}\pi/\mathbb{Z}\pi)$ and M is a $\mathbb{Z}\pi$-module (see (2.2)).

We now use conglomerates to carry out a more complete discussion of linking and self-linking numbers than was possible in Chapter 4A. Let $\mathcal{C}^{k+1}_\mu = \bigcup \mathfrak{M}(i)$ be a conglomerate constructed according to the $(n\times n)$-matrix $\mu \in M_n(\mathbb{Z}\pi, \mathbb{Q}\pi)^\times$, let N^{2k+1} be a p.l. manifold with $\pi_1 N = \pi$ and $F: \mathcal{C}^{k+1}_\mu \to N^{2k+1}$ a framed immersion. Using the technique of [W1, §5] or (4A.37), we may define a self-linking form $\psi: \mathrm{im}\{H_k(\mathcal{C}^{k+1}_\mu) \to H_k(N^{2k+1})\} \to \mathbb{Q}\pi/J_\epsilon$, $\epsilon = (-1)^{k+1}$. Let $b_i\mathcal{C}^{k+1}_\mu$ denote the ith component of the bockstein (see (4B.3)) and let F' denote F pushed a non-zero distance in the direction of the first vector of its complementary framing. Recall from (2.19) the notion of "covering" and from (2.5d) that of "ϵ-linking form."

4B.33 Proposition. Let $F: \mathcal{C}^{k+1}_\mu \to N^{2k+1}$ be a framed immersion as above, such that $F(\mathring{\mathcal{C}}^{k+1}_\mu)$ intersects $F(b\mathcal{C}^{k+1}_\mu)$ transversely in isolated points, and $F_*: H_k(\mathcal{C}^{k+1}_\mu) \to H_k(N^{2k+1})$ is injective. Then there is a unique $(n\times n)$-matrix τ with entries in $\mathbb{Q}\pi$ such that

4B.34 $$(\tau\mu)_{ij} = F'(b_i\mathcal{C}_\mu^{k+1}) \pitchfork F(\mathfrak{m}(j)).$$

If R denotes the resolution corresponding to $\tilde{\mathcal{C}}_\mu^{k+1}$ (see (4B.9)), the pair (R,τ) covers the ϵ-linking form $(F_*H_k(\mathcal{C}_\mu^{k+1}),\varphi,\psi)$, where

$$\varphi: F_*H_k(\mathcal{C}_\mu^{k+1}) \times F_*H_k(\mathcal{C}_\mu^{k+1}) \longrightarrow \mathbb{Q}\pi/\mathbb{Z}\pi$$

and

$$\psi: F_*H_k(\mathcal{C}_\mu^{k+1}) \longrightarrow \mathbb{Q}\pi/J_\epsilon$$

are linking and self-linking forms, respectively, and $\epsilon = (-1)^{k+1}$.

<u>Proof</u>. For the proof let $\mathcal{C}_\mu = F(\mathcal{C}_\mu^{k+1})$, $\mathfrak{m}(i) = F(\mathfrak{m}(i))$, $b_i = b_i\mathcal{C}_\mu^{k+1}$, and $b_i' = F'(b_i)$. Let σ be the $(n\times n)$-matrix with $\sigma_{ij} = b_i' \pitchfork \mathfrak{m}(j)$, and set $\tau = \sigma\mu^{-1}$. This makes sense because μ is $\mathbb{Q}\pi$-invertible. For the same reason there is $t \in \mathbb{Z}$ and an $(n\times n)$-matrix ν with entries in $\mathbb{Z}\pi$ such that $\mu\nu = tI_n$. By construction (4B.1), there is an equation of chains, $\partial\mathfrak{m}(\ell) = \sum_i b_i\mu_{i\ell}$. Thus, $\partial(\sum_\ell \mathfrak{m}(\ell)\nu_{\ell j}) = \sum_{i,\ell} b_i\mu_{i\ell}\nu_{\ell j} = tb_j$. Let $[b_i]$ denote the homology class of b_i. Then

$$\begin{aligned}\varphi([b_i],[b_j]) &\equiv \frac{1}{t}(b_i' \pitchfork \sum_\ell \mathfrak{m}(\ell)\nu_{\ell j}) \\ &\equiv \frac{1}{t}(\sum_\ell (\tau\mu)_{i\ell}\nu_{\ell j}) \\ &\equiv \frac{1}{t}(\tau_{ij}t) \equiv \tau_{ij} \pmod{\mathbb{Z}\pi}.\end{aligned}$$

The same argument with $i = j$ shows $\psi([b_i]) \equiv \tau_{ii} \pmod{J_\epsilon}$, $\epsilon = (-1)^{k+1}$. This completes the proof.

For future reference, we state a useful lemma contained in the preceding proof.

4B.35 <u>Lemma</u>. In the notation of (4B.33), if t and μ are such that $\mu\nu = tI_n$, then we have the equation of chains:

$$\partial(\sum_\ell \mathfrak{m}(\ell)\nu_{\ell j}) = t(b_j\mathcal{C}_\mu^{k+1}).$$

Here is a partial converse to (4B.33).

4B.36 Proposition. Keeping the assumptions of (4B.33), suppose $\varphi(S \times S) \equiv 0 \equiv \psi(S)$ where $S = F_* H_k(\mathcal{C}_\mu^{k+1})$ and φ and ψ are the linking and self-linking forms, respectively. Then F is regularly homotopic to $G: \mathcal{C}_\mu^{k+1} \to N^{2k+1}$ where $G(b\mathcal{C}_\mu^{k+1}) \cap G(\mathring{\mathcal{C}}_\mu^{k+1}) = \emptyset$.

Proof. We keep the abbreviated notation used in the proof of (4B.33). By assumption and (4B.35), for each $i,j = 1,\ldots,n$, and $\epsilon = (-1)^{k+1}$

4B.37 $$\lambda_{ij} = b_i' \pitchfork \left(\sum_\ell \mathfrak{m}(\ell)\nu_{\ell j}\right) \in \begin{cases} t\mathbb{Z}\pi, & j \neq i \\ tJ_\epsilon, & j = i, \end{cases}$$

for some $t \in \mathbb{Z}$ and some $(n\times n)$-matrix ν for which $\mu\nu = tI_n$. Let $\sigma_{i\ell} = b_i' \pitchfork \mathfrak{m}(\ell)$. Then $\lambda_{ij} = \sum_\ell \sigma_{i\ell}\nu_{\ell j}$, so $\lambda = \sigma\nu$. It suffices to show σ is the zero matrix, for then we can use the Whitney device (on the manifolds b_i' and $\mathring{\mathfrak{m}}(\ell)$) to modify $F|b\mathcal{C}_\mu^{k+1}$ to make its image disjoint from $F(\mathring{\mathcal{C}}_\mu^{k+1})$. To show $\sigma = 0$, it suffices to show $\lambda = 0$, since ν is $\mathbb{Q}\pi$-invertible.

The idea of proof is to reduce to the situation of (4A.39). Fix j and consider the chain $\mathfrak{n}(j) = \sum_\ell \mathfrak{m}(\ell)\nu_{\ell j}$. By lemma (4B.35),

4B.38 $$\partial(\mathfrak{n}(j)) = tb_j,$$

so the sheets of $\mathfrak{n}(j)$ abutting to b_ℓ, $\ell \neq j$, can be arranged in pairs corresponding to the same $g \in \pi$ but opposite orientation at b_ℓ. Since F may be assumed to be an imbedding near $b\mathcal{C}_\mu^{k+1}$ (general position), the neighborhoods of b_ℓ containing each such pair of sheets may be pushed a different distance in the direction of the second vector of the complementary framing, so that the singularity of $\mathfrak{n}(j)$ at b_ℓ is replaced by singularities of the form $C(2) \times S^k$ where $C(2)$ is the cone on two points:

4B.39 Figure.

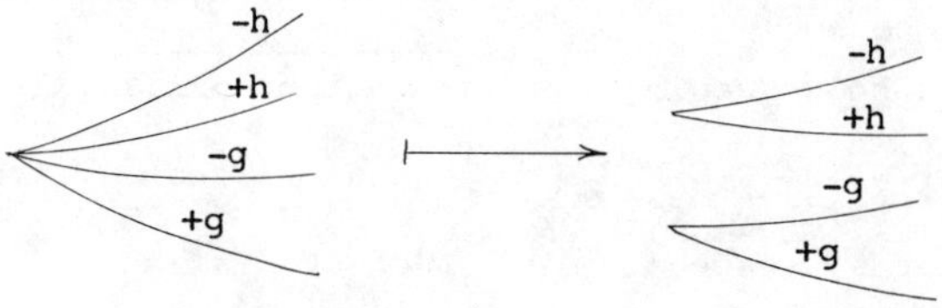

Straightening out the cone point,

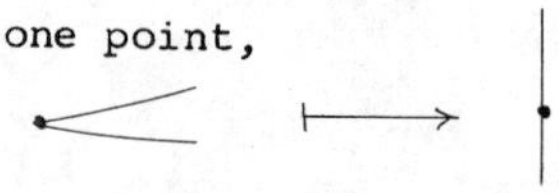

converts $C(2) \times S^k$ to $I \times S^k$, I an interval, thus removing all singularities of $\mathfrak{n}(j)$ at b_ℓ, $\ell \neq j$. If $\ell = j$, the same procedure may be followed at $b_\ell = b_j$, leaving only a singularity of type $C(t) \times S^k$, corresponding to tb_j in (4B.38).

Let $\hat{\mathfrak{n}}(j)$ denote the result of these modifications of $\mathfrak{n}(j)$. Then $\hat{\mathfrak{n}}(j)$ is a manifold except at b_j, where it has the local structure of $b\mathfrak{m}_t^{k+1}$. In particular, we may now apply the technique of (4A.39) using the assumption (4B.37) to bring about $b_i' \cap \hat{\mathfrak{n}}(j) = \emptyset$, for all i and j. The conversion of $\mathfrak{n}(j)$ to $\hat{\mathfrak{n}}(j)$ was purely local and is clearly reversible. Thus, by a regular homotopy of F, we may take $\lambda_{ij} = 0$ in (4B.37), as required.

We can use the above results to give a description of linking and self-linking of cycles in the total space of the "trivial bundle" over $\widetilde{\mathcal{C}}_\mu^k$, which is analogous to that of intersection and self-intersection of cycles in the total space of the trivial sphere bundle over S^k. Recall from (4B.18) that $n\widetilde{\mathcal{C}}_\mu^k \times D^{k-1}$ admits a free π-action.

4B.40 Proposition. Let $\widetilde{\mathcal{C}}_\mu^k$ be constructed according to the matrix $\mu \in M_n(\mathbb{Z}\pi, \mathbb{Q}\pi)^\times$, and let $\widetilde{U}$ denote $n\widetilde{\mathcal{C}}_\mu^k \times D^{k-1}$. The sequence of $\mathbb{Z}\pi$-modules

4B.41 $$H_k(\widetilde{U}, \partial\widetilde{U}) \rightarrowtail H_{k-1}(\partial\widetilde{U}) \twoheadrightarrow H_{k-1}(\widetilde{U})$$

is short exact and is split by

4B.42 $$s\colon H_{k-1}(\widetilde{U}) \rightarrowtail H_{k-1}(\partial\widetilde{U})$$

induced by pushing $\tilde{\mathcal{C}}^k_\mu \subset \mathring{\tilde{U}}$ out to $\partial\tilde{U}$ by any vector of the complementary framing (or D^{k-1}-direction). The inclusion δ of the dual Moore space (4B.31) into U induces an isomorphism

$$\text{4B.43} \qquad \delta_*\colon H_*(\tilde{\tilde{\mathcal{C}}}^{k+1}_{\bar\mu}, \tilde{\mathcal{C}}^k_{\bar\mu}) \xrightarrow{\ \cong\ } H_*(\tilde{U}, \partial\tilde{U}).$$

Proof. (4B.41) is part of the homology exact sequence of the pair $(\tilde{U}, \partial\tilde{U})$. Since $\tilde{\mathcal{C}}^k_\mu \simeq \tilde{U}$, $H_k(\tilde{U}) = 0$ by (4B.4(ii)); and $H_{k-1}(\tilde{U},\partial\tilde{U}) \cong H_k(\tilde{U})^\wedge$ by duality, so (4B.41) is short exact. (Recall from (2.2) that $H_k(\tilde{U})^\wedge$ means $\mathrm{Hom}_{\mathbb{Z}\pi}(H_k(\tilde{U}), \mathbb{Q}\pi/\mathbb{Z}\pi)$.) That s splits $H_{k-1}(\partial\tilde{U}) \twoheadrightarrow H_{k-1}(\tilde{U})$ is clear.

To show (4B.43) is an isomorphism it suffices to show it is surjective for $* = k$ because the modules involved are finite and abstractly isomorphic: $H_k(\tilde{U},\partial\tilde{U}) \cong H_{k-1}(\tilde{U})^\wedge$ by duality, and $H_k(\tilde{\tilde{\mathcal{C}}}^{k+1}_{\bar\mu}, \tilde{\mathcal{C}}^k_{\bar\mu}) \cong H_{k-1}(\tilde{\mathcal{C}}^k_{\bar\mu}) \cong H_{k-1}(\tilde{\mathcal{C}}^k_\mu)^\wedge \cong H_{k-1}(\tilde{U})^\wedge$, by (4B.25(ii)) and (4B.32); for $* \neq k$, $H_*(\tilde{U},\partial\tilde{U}) \cong H_{*-1}(\tilde{U})^\wedge = 0$ and $H_*(\tilde{\tilde{\mathcal{C}}}^{k+1}_{\bar\mu}, \tilde{\mathcal{C}}^k_{\bar\mu}) = 0$, by (4B.25). From the surjectivity of j" in (4B.4) and (4B.25(i)) it follows that $H_k(b\tilde{\tilde{\mathcal{C}}}^{k+1}_{\bar\mu}, b\tilde{\tilde{\mathcal{C}}}^k_{\bar\mu}) \to H_k(\tilde{\tilde{\mathcal{C}}}^{k+1}_{\bar\mu}, \tilde{\mathcal{C}}^k_{\bar\mu})$ is surjective. Thus to prove δ_k is surjective it suffices to show that the relative homology classes $\{[{}_*D^k_{e,i}] \mid i=1,\ldots,n\}$ generate $H_k(\tilde{U},\partial\tilde{U})$ where $b\tilde{\tilde{\mathcal{C}}}^{k+1}_{\bar\mu} = \bigcup_{g,i} {}_*D^k_{g,i}$ (see (4B.31)).

Let $H_k(\tilde{\mathcal{C}}^k_\mu, b\tilde{\mathcal{C}}^k_\mu) \overset{\mu}{\rightarrowtail} H_{k-1}(b\tilde{\mathcal{C}}^k_\mu) \overset{j}{\twoheadrightarrow} H_{k-1}(\tilde{\mathcal{C}}^k_\mu) = H_{k-1}(\tilde{U})$ be the resolution corresponding to $\tilde{\mathcal{C}}^k_\mu$ in Definition (4B.9), with bases $\{e_i\}$ for $H_k(\tilde{\mathcal{C}}^k_\mu, b\tilde{\mathcal{C}}^k_\mu)$ and $\{[S^{k-1}_{e,j}]\} = \{f_j\}$ for $H_{k-1}(b\tilde{\mathcal{C}}^k_\mu)$ as chosen there. Let $\overline{H_{k-1}(b\tilde{\mathcal{C}}^k_\mu)} \overset{\bar\mu}{\rightarrowtail} \overline{H_k(\tilde{\mathcal{C}}^k_\mu, b\tilde{\mathcal{C}}^k_\mu)} \overset{\bar{j}}{\twoheadrightarrow} H_{k-1}(\tilde{U})^\wedge$ be its corresponding dual resolution (2.7). If

$$\alpha\colon H_k(\tilde{U},\partial\tilde{U}) \times H_{k-1}(\tilde{U}) \longrightarrow \mathbb{Q}\pi/\mathbb{Z}\pi$$

is the duality (linking) pairing, it is enough by nonsingularity of α to show, for all $i = 1,\ldots,n$,

$$\text{4B.44} \qquad \mathrm{Ad}(\alpha)[{}_*D^k_{e,i}] = \bar{j}(\bar{e}_i)$$

where $\{\bar{e}_i\}$ is the dual basis for $\overline{H_{k-1}(\tilde{\mathcal{C}}^k_\mu, b\tilde{\mathcal{C}}^k_\mu)}$. By (2.17), $(\mathfrak{J}(\bar{e}_i))(j(f_\ell)) \equiv (\mu^{-1})_{i\ell} \pmod{\mathbb{Z}\pi}$. By definition, $j(f_\ell) = [S^{k-1}_{e,\ell}]$, so

$$(\mathrm{Ad}(\alpha)[{}_*D^k_{e,i}])(j(f_\ell)) = \alpha([{}_*D^k_{e,i}],[S^{k-1}_{e,\ell}])$$

$$\equiv \frac{1}{t}({}_*D^k_{e,i} \pitchfork (\sum \mathfrak{M}(e,j)\nu_{j\ell})) \pmod{\mathbb{Z}\pi} \quad \text{by (4B.35)}$$

$$\equiv \frac{1}{t}\nu_{i\ell} \pmod{\mathbb{Z}\pi}, \quad \text{since } {}_*D^k_{e,i} \pitchfork \mathfrak{M}(e,j)$$

$$= \delta_{ij} \text{ (Kronecker delta) by construction of } \tilde{\mathcal{C}}^{k+1}_{\bar{\mu}}$$

$$\equiv (\mu^{-1})_{i\ell} \pmod{\mathbb{Z}\pi}, \text{ since } \mu\nu = tI_n \quad (4B.35).$$

Thus $(\mathrm{Ad}(\alpha)[{}_*D^k_{e,i}])(j(f_\ell)) = \mathfrak{J}(\bar{e}_i)(j(f_\ell))$ which proves (4B.44), since $\{j(f_\ell) \mid \ell = 1,\ldots,n\}$ generates $H_{k-1}(\tilde{U})$. This completes the proof of (4B.40).

4B.45 Proposition. Keeping the notation of (4B.40), the $(-1)^k$-form $(H_{k-1}(\partial\tilde{U}),\varphi,\psi)$, furnished by linking (φ) and self-linking (ψ) of cycles in $\partial\tilde{U}$, is hyperbolic with complementary summands given by the images of the injection $H_k(\tilde{U},\partial\tilde{U}) \rightarrowtail H_{k-1}(\partial\tilde{U})$ and the section $s: H_{k-1}(\tilde{U}) \rightarrowtail H_{k-1}(\partial\tilde{U})$.

Proof. Since each summand in question is generated by the bocksteins of a framed imbedded conglomerate, it is totally isotropic by (4B.33). Using $\mathrm{Ad}(\alpha): H_k(\tilde{U},\partial\tilde{U}) \xrightarrow{\cong} H_{k-1}(\tilde{U})^\wedge$ to identify $H_k(\tilde{U},\partial\tilde{U})$ with $H_{k-1}(\tilde{U})^\wedge$, we have identified $H_{k-1}(\partial\tilde{U})$ with $H_{k-1}(\tilde{U})^\wedge + H_{k-1}(\tilde{U})$. It remains therefore (by Definition (2.11)(b)) to verify that

$$\varphi((\mathrm{Ad}(\alpha))^{-1} \times 1): H_{k-1}(\tilde{U})^\wedge \times H_{k-1}(\tilde{U}) \longrightarrow \mathbb{Q}\pi/\mathbb{Z}\pi$$

is the natural form (2.6)(c). This follows from naturality properties of φ and α.

We will now determine the relation between conglomerates and surgery.

Roughly speaking, the following pair of propositions shows there is a one-to-one correspondence between framed imbedded conglomerates in a manifold M and bordisms W of M having handles in two adjacent dimensions.

4B.46 <u>Proposition</u>. Let M^m be a p.l. manifold and μ a matrix in $M_n(\mathbb{Z}\pi,\mathbb{Q}\pi)^{\times}$, where $\pi = \pi_1 M^m$. A framed imbedded conglomerate $\mathcal{C}_\mu^k \subset M^m$, $k < m-1$, determines a bordism W^{m+1} of M^m such that

(i) (W^{m+1},M^m) has handles of index k and $k+1$;

(ii) there is a p.l. homeomorphism $W^{m+1} \approx (M^m \times I) \cup (n\mathcal{C}_\mu^{-k+1} \times D^{m-k-1})$, where the union is taken along $n\mathcal{C}_\mu^k \times D^{m-k-1} \subset n\mathcal{C}_\mu^{-k+1} \times D^{m-k-1}$ and $n\mathcal{C}_\mu^k \times D^{m-k-1} \subseteq M^m \times 1$, the neighborhood of $\mathcal{C}_\mu^k$ in M^m afforded by the framing; and

(iii) $$H_i(W^{m+1},M^m) \cong \begin{cases} H_k(\tilde{\mathcal{C}}_\mu^{k+1},\tilde{\mathcal{C}}_\mu^k) = H_{k-1}(\tilde{\mathcal{C}}_\mu^k), & i = k \\ 0, & i \neq k. \end{cases}$$

<u>Proof</u>. Let $b_i\mathcal{C}_\mu^k$ denote the ith bockstein of $\mathcal{C}_\mu^k$, $i = 1,\ldots,n$. Since $\mathcal{C}_\mu^k$ is framed and imbedded, there is, for each i, a neighborhood U_i of $b_i\mathcal{C}_\mu^k$ in M^m and homeomorphisms $\alpha_i\colon U_i \overset{\approx}{\to} S^{k-1} \times D^{m-k+1}$ and $\beta_i\colon U_i \cap \mathfrak{M}(1) \overset{\approx}{\to} S^{k-1} \times C(r_i)$ where $\beta_i(b_i\mathcal{C}_\mu^k) = S^{k-1}$, $\mathcal{C}_\mu^k = \cup\mathfrak{M}(i)$, and $C(r_i)$ is the cone on r_i points (4A.4), such that the diagram

4B.47 <u>Diagram</u>.
$$\begin{array}{ccc} U_i \cap \mathfrak{M}(1) & \hookrightarrow & U_i \\ \approx\downarrow \beta_i & & \approx\downarrow \alpha_i \\ S^{k-1}\times C(r_i) & \xrightarrow{I\times J} & S^{k-1}\times D^{m-k+1} \end{array}$$
commutes.

In (4B.47), $I\colon S^{k-1} \to S^{k-1}$ is the identity and $J = i\circ c$, where $c\colon C(r_i) \to D^2$ is given in (4A.5) and $i\colon D^2 \to D^{m-k+1}$ is inclusion into the first two factors of D^{m-k+1}. The last $(m-k-1)$ factors of D^{m-k+1} present the complementary framing of $\mathfrak{M}(1)$ in a neighborhood of $b_i\mathcal{C}_\mu^k$. Figure (4B.48) is a cross section $s_0 \times D^{m-k+1}$ of $S^{k-1} \times D^{m-k+1}$ at some $s_0 \in S^{k-1}$.

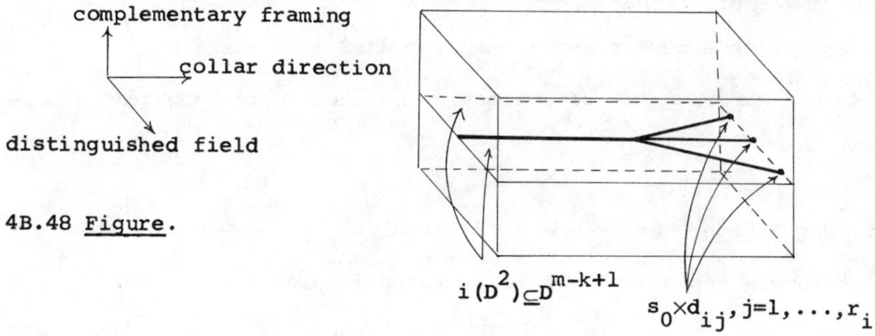

4B.48 <u>Figure</u>.

Let $\beta_i(\partial U_i \cap \mathfrak{M}(1)) = \bigcup_{j=1}^{r_i} S^{k-1} \times d_{ij}$, where the d_{ij} are the right end-points of $C(r_i)$ (the z_i in Figure (4A.6)).

Surgery on $b_i \mathcal{C}_\mu^k$ removes $\mathring{U}_i$ and replaces it with $H_i = D_i^k \times S^{m-k}$. Let $\bar{H}_i = D_i^k \times D^{m-k+1}$; then $V^{m+1} = M^m \times I \cup (\bigcup_i \bar{H}_i)$ is the bordism for surgery on the $b_i \mathcal{C}_\mu^k$, where the union is taken over $\cup U_i$ in $M^m \times \{1\}$ and $\cup(\partial D_i^k \times D^{m-k+1})$ in $\cup \bar{H}_i$. $\mathfrak{M}(1) - \bigcup_i (U_i \cap \mathfrak{M}(1))$ is a punctured sphere with boundary $\bigcup_{i=1}^{n} \bigcup_{j=1}^{r_i} \beta_i^{-1}(S^{k-1} \times d_{ij})$ and $\partial(D_i^k \times J(d_{ij})) = \beta_i^{-1}(S^{k-1} \times d_{ij})$ so

$$(\mathfrak{M}(1) - \bigcup_i (U_i \cap \mathfrak{M}(1)) \cup (\bigcup_{i,j} D_i^k \times J(d_{ij}))$$

is the image of a framed imbedding $g_i: S^k \to M'^m$, where $\partial V^{m+1} = M^m \cup M'^m$. See Figure (4B.49)

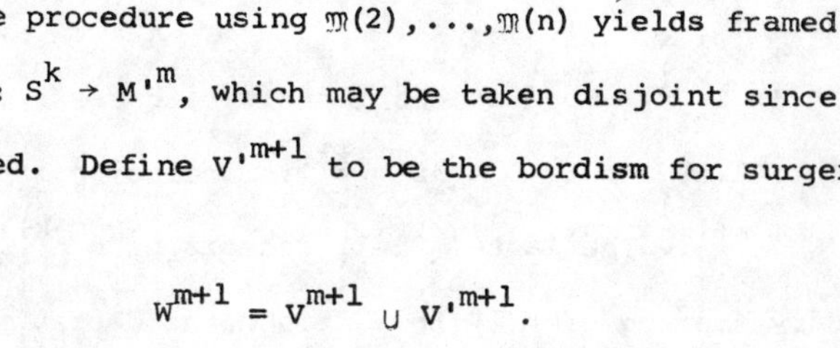

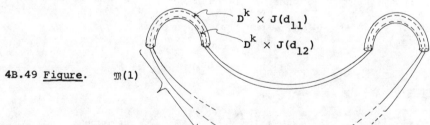

4B.49 <u>Figure</u>.

Carrying out the same procedure using $\mathfrak{M}(2),\dots,\mathfrak{M}(n)$ yields framed imbeddings $g_2,\dots,g_n: S^k \to M'^m$, which may be taken disjoint since $\mathcal{C}_\mu^k = \cup\mathfrak{M}(i)$ is imbedded. Define V'^{m+1} to be the bordism for surgery on the $g_i(S^k)$ and let

$$W^{m+1} = V^{m+1} \cup V'^{m+1}.$$

Observe that the (framed) canonical imbedding

$I \times J\colon S^{k-1} \times C(r_i) \to S^{k-1} \times D^{m-k+1}$ in (4B.47) can be extended to a (framed) imbedding $I_D \times J\colon D^k \times C(r_i) \to D^k_i \times D^{m-k+1} = \bar{H}_i$, where $I_D\colon D^k \to D^k_i$ is the "identity",

4B.50 Figure.

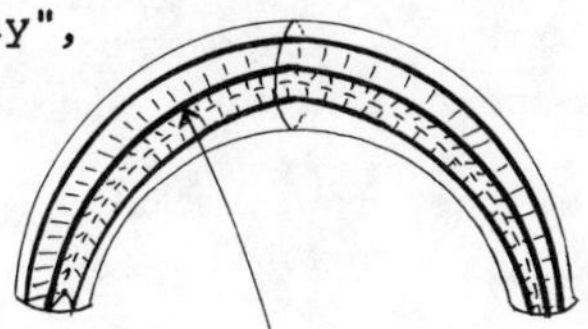

$(I_D \times J)(D^k \times C(r_i))$

and

4B.51 $$(I_D \times J)\Big(\bigcup_{i,j} D^k \times J(d_{ij})\Big) = g_1(S^k) \cap \Big(\bigcup_i H_i\Big);$$

the analogous equation holds for $g_2,\ldots,g_n$. Thus, if $B^{k+1}_1,\ldots,B^{k+1}_n \subset V'^{m+1}$ denote the core discs attached to M'^m for surgery on $g_1(S^k),\ldots,g_n(S^k)$,

4B.52 $$\Big(\bigcup_i (I_D \times J)(D^k \times C(r_i))\Big) \cup \Big(\bigcup_j B^{k+1}_j\Big)$$

is a framed imbedded conglomerate $\mathcal{C}^{-k+1}_\mu \subset V^{m+1} \cup V'^{m+1} = W^{m+1}$. The union in (4B.52) must be taken so as to give all r_i corresponding to the d_{ij} on the left side of (4B.51) where we carry out the above procedure for $g_2,\ldots,g_n$ in place of g_1. It is left to the reader to show that (4B.46) (ii) is satisfied. Part (iii) follows from (ii), excision, and (4B.25).

4B.53 <u>Definition</u>. In the notation of (4B.46), the construction of W^{m+1} from the framed imbedding $F\colon\mathcal{C}^k_\mu \to M^m$ is called <u>surgery on $\mathcal{C}^k_\mu$</u>.

4B.54 <u>Proposition</u>. Let M^m be a p.l. manifold and let W^{m+1} be a bordism of M^m with handles of index k and $k+1$, for some k, $2 < k < m-2$, such that $H_*(W^{m+1},M^m)\otimes_{\mathbb{Z}\pi} \mathbb{Q}\pi \equiv 0$, $\pi = \pi_1 M^m$. Then there is a matrix $\mu \in M_n(\mathbb{Z}\pi,\mathbb{Q}\pi)^\times$ and a framed imbedding $F\colon \mathcal{C}^k_\mu \to M^m$ such that W^{m+1} is the bordism of M^m given by surgery on $F(\mathcal{C}^k_\mu)$ in (4B.46). In particular, condition (4B.46)(ii) is satisfied.

Proof. Arrange the handle addition so that $W^{m+1} = V^{m+1} \cup V'^{m+1}$, $V^{m+1} = M^m \times I \cup (\bigcup_i \bar{H}_i)$ where each $\bar{H}_i$ is a k-handle, $\bar{H}_i = D^k_i \times I^{m-k+1}$, $\partial V^{m+1} = M^m \cup M'^m$; $V'^{m+1} = M'^m \times I \cup (\bigcup_j \bar{K}_j)$, where for each j, $\bar{K}_j = D^{k+1}_j \times I^{m-k}$, a (k+1)-handle. (Here I^ℓ is the ℓ-fold product of $I = [-1,1]$.) For the rest of the proof we omit superscripts indicating dimension of the manifolds W^{m+1}, M^m, etc.

Since W is constructed by adding k- and (k+1)-handles to M, the only non-zero groups $H_*(W,M)$ appear in the exact sequence

$$H_{k+1}(W,M) \rightarrowtail H_{k+1}(V',M') \xrightarrow{\partial} H_k(V,M) \twoheadrightarrow H_k(W,M).$$

Since $H_{k+1}(V',M')$ has no $\mathbb{Z}$-torsion (it is $\mathbb{Z}\pi$-free) neither does $H_{k+1}(W,M)$; on the other hand, the latter is by hypothesis a $\mathbb{Z}$-torsion $\mathbb{Z}\pi$-module. Hence it is zero, so the remaining short exact sequence

4B.55 $$H_{k+1}(V',M') \overset{\partial}{\rightarrowtail} H_k(V,M) \twoheadrightarrow H_k(W,M)$$

exhibits $H_k(W,M)$ as an element of $\mathfrak{D}^1_F$ (2.7), since $H_{k+1}(V',M')$ and $H_k(V,M)$ are $\mathbb{Z}\pi$-free. The latter modules have $\mathbb{Z}\pi$-bases of the same cardinality, say, n, since $\partial \otimes \mathbb{Q}\pi$ is an isomorphism. There are thus n $\bar{K}_j$'s and n $\bar{H}_i$'s.

For each $i = 1,\dots,n$, let $f_i\colon S^{k-1} \times I^{m-k+1} \to M$ be the attaching map for $\bar{H}_i = D^k_i \times I^{m-k+1}$, and set $H_i = D^k_i \times \partial(I^{m-k+1})$, so that $M' = (M - \bigcup_{i=1}^n f_i(S^{k-1} \times \mathring{I}^{m-k+1}) \cup (\bigcup_{i=1}^n H_i)$. Let $g_i\colon S^k \times I^{m-k} \to M'$ be the attaching map for $\bar{K}_i = D^{k+1}_i \times I^{m-k}$. Let $F^*_i = 0_i \times I^{m-k+1} \subset \bar{H}_i$ (F^*_i is the dual cell), where $0_i \in D^k_i$ is the center of D^k_i. Arrange by an isotopy that each $g_i|S^k \times \vec{0}$ meet ∂F^*_j transversely in isolated points, where $\vec{0} = (0,\dots,0) \in I^{m-k+1}$. By the ambient isotopy theorem, this does not change (W,M) up to homeomorphism.

Let $F_i = D^k_i \times \vec{0} \subseteq D^k_i \times I^{m-k+1} = \bar{H}_i$ and $E_j = D^{k+1}_j \times \vec{0} \subseteq D^{k+1}_j \times I^{m-k} = \bar{K}_j$. Then the $[E_j]$ form a basis for $H_{k+1}(V',M')$ and the $[F_i]$ for

$H_k(V,M)$, where $[X]$ denotes the homology class of the cycle X. Expressing ∂ in (4B.55) as a matrix μ in terms of these bases, we have the algebraic chain intersection

4B.56 $$\partial E_i \pitchfork F_j^* = \mu_{ij}.$$

For some i and j, let $\mu_{ij} = \Sigma\, n_h h$. By the Whitney device we may construct an isotopy of $g_i|S^k \times \vec{0}$ so that there are exactly n_h points of intersection of $g_i(S^k) = \partial E_i$ with F_j^* corresponding to each $h \in \pi$.

4B.57 Figure.

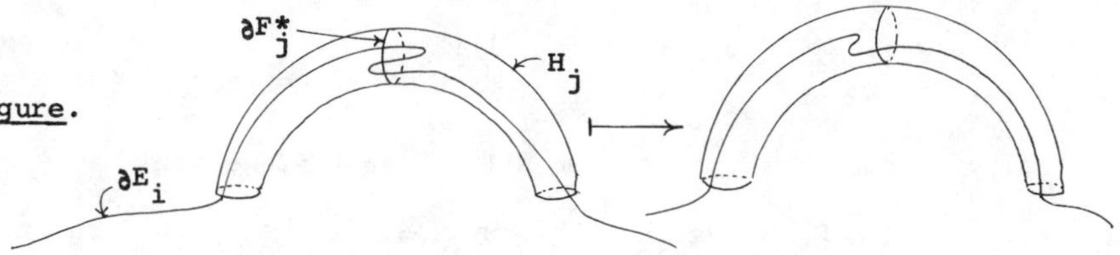

Now fix i and j and consider the intersection of $\partial E_i = g_i(S^k \times \vec{0})$ with the handle $H_j = D_j^k \times \partial(I^{m-k+1})$. Express the dual cell F_j^* as $0_i \times (I^2 \times I^{m-k-1})$ and let $c\colon C(r_{ij}) \to I^2$ be the imbedding of the cone on r_{ij} points constructed in (4A.5), where $r_{ij} = \Sigma_g |n_g|$, $\mu_{ij} = \Sigma_g\, n_g g$; let $\{x_1,\ldots,x_{r_{ij}}\}$ denote the set of end points of the cone (see Fig. (4A.4)). If T_{ij} denotes the point set intersection $\partial E_i \cap F_j^*$, we can clearly arrange that

$$T_{ij} = \{0_i \times z_1 \times \vec{0},\ldots,0_i \times z_{r_{ij}} \times \vec{0}\}$$

where $z_\ell = c(x_\ell)$, 0_i is the center of D_i^k and $\vec{0} = (0,\ldots,0) \in I^{m-k-1}$. (See Fig. (4B.48).) Thus, at each point of T_{ij} there is a one frame field ϵ' coming from $\nu_{C(r_{ij})}$ (4A.5) in the I^2-factor of F_j^* and a complementary $(m-k-1)$-frame field η' coming from the I^{m-k-1}-factor. Together, ϵ' and η' are a complementary pair of trivial subbundles furnishing a framing of $\nu((\cup_\ell D_j^k \times z_\ell \times \vec{0}) \subset H_j)|_{T_{ij}}$ which can clearly be arranged to agree with the natural framing of $\nu(g_i(S^k \times \vec{0}) \subset g_i(S^k \times I^{m-k}))|_{T_{ij}}$, where $\nu(X \subset Y)|_Z$ denotes the normal bundle of X in Y, restricted to a subset Z of X. (The framings can be arranged to agree because the second framing extends over the disc F_i.)

Clearly ϵ' and η' extend to normal fields ϵ and η, normal to $D_j^k \times c(C(r_{ij})) \times \vec{0}$ in $\bar{H}_j$. We can now "spread" the above framing data over the discs $D_j^k \times z_\ell \times 0$, $\ell = 1,\dots,r_{ij}$. Namely, by an isotopy we can clearly arrange that

(i) $g_i(S^k \times \vec{0}) \cap H_j = \bigcup_\ell (D_j^k \times z_\ell \times \vec{0})$ and

(ii) the framing of $\nu(g_i(S^k \times \vec{0}) \subset H_j)$ given by ϵ and η agrees with the natural framing of

$$\nu(g_i(S^k \times \vec{0}) \subset g_i(S^k \times I^{m-k}))|_{g_i(S^k \times \vec{0}) \cap H_j}.$$

Carrying out this process for all i and j, we set $\bar{\mathfrak{M}}(i) = E_i^{k+1} \cup (\bigcup_{j=1}^n D_j^k \times C(r_{ij}) \times \vec{0})$. By (4B.56) $\bigcup_i \bar{\mathfrak{M}}(i)$ is a relative conglomerate $\bar{\mathcal{C}}_\mu^{k+1} \subset W$, constructed according to the matrix μ. $\bar{\mathcal{C}}_\mu^{k+1}$ is framed by condition (ii) above and the fact that ϵ' and η' extend over $D_j^k \times c(C(r_{ij})) \times \vec{0}$ (to give the distinguished and complementary framings ϵ and η in a neighborhood of $b\bar{\mathcal{C}}_\mu^{k+1} = \bigcup_j E_j$). The boundary, $\mathcal{C}_\mu^k$, is framed imbedded in M and (4B.46(ii)) is clearly satisfied. This completes the proof.

The next proposition shows there are framed imbedded conglomerates in essentially all dimensions; however, those constructed are homologically trivial.

4B.58 Proposition. Let N be a p.l. manifold and let $\mu \in M_n(\mathbb{Z}\pi, \mathbb{Q}\pi)^\times$. If $k \leq \dim N - 2$, there is a framed imbedding

$$\bar{F}: \bar{\mathcal{C}}_\mu^{k+1} \longrightarrow N.$$

Proof. Choose a base point x_0 and a $(k+1)$-disc $D^{k+1}(x_0)$ containing it. Choose a disc $B \subseteq N$ of dimension $\dim N$, $B \cap D^{k+1}(x_0) = \emptyset$, and a system of paths $\{\ell_g | g \in \pi\}$ from x_0 to B furnishing the $|\pi|$ liftings of

B to $\widetilde{N}$, the universal cover of N. Recalling the construction (4B.20) of $\bar{\mathcal{C}}^{k+1}_{\mu}$, let $\mu_{11} = \Sigma\, n_g g$ and let $\bar{F}_g : \bar{\mathfrak{M}}^{k+1}_{|n_g|} \to B$, $g \in \pi$, be framed imbeddings (see (4A.12)). Since $\bar{\mathfrak{M}}^{k+1}_{|n_g|}$ is essentially 1-dimensional (3.9), the $\bar{F}_g$ may be assumed disjoint. Using a "tunnel" $D^k \times I$ running along ℓ_g, take boundary-connected-sum of each $\bar{F}_g(\bar{\mathfrak{M}}^{k+1}_{|n_g|})$ with $D^{k+1}(x_0)$ (where $\bar{F}_g(\mathfrak{M}^{k}_{|n_g|})$ is the "boundary" of $\bar{F}_g(\bar{\mathfrak{M}}^{k+1}_{|n_g|})$). The result is the image of a framed imbedding $\bar{F}_{11}: \bar{\mathfrak{M}}^{k+1}_{\{|n_g|\,|\,g\in\pi\}} \to N$. Continuing in this way, the rest of the construction of $\bar{\mathcal{C}}^{k+1}_{\mu}$ can be carried out in N, yielding $\bar{F}$. Once again there is no problem with intersections since $\bar{\mathcal{C}}^{k+1}_{\mu}$ is essentially 1-dimensional by (4B.25(i)).

4B.59 <u>Definition</u>. A mapping $F: \mathcal{C}^{k}_{\mu} \to N$ is called <u>trivial</u> if it extends to a mapping $\bar{F}: \bar{\mathcal{C}}^{k+1}_{\mu} \to N$. By (4B.58), trivial framed imbeddings $F: \mathcal{C}^{k}_{\mu} \to N$ exist.

The final proposition of Chapter 4B uses the machinery developed to show how homology and linking forms in an odd-dimensional manifold are affected by surgery on a trivially imbedded conglomerate.

4B.60 <u>Proposition</u>. Let N^{2k+1} be a p.l. manifold, $\mu \in M_n(\mathbb{Z}\pi, \mathbb{Q}\pi)^{\times}$, $\pi = \pi_1 N$, and let $\bar{F}: \bar{\mathcal{C}}^{k+1}_{\mu} \to N^{2k+1}$ be a framed immersion such that $\bar{F}|\mathcal{C}^{k}_{\mu}$ is an imbedding and $\bar{F}(\dot{\bar{\mathcal{C}}}^{k+1}_{\mu}) \cap F(b\bar{\mathcal{C}}^{k+1}_{\mu}) = \emptyset$. If W^{2k+2} is the bordism for surgery on $\bar{F}(\mathcal{C}^{k}_{\mu})$ and $\partial W^{2k+2} = N^{2k+1} \cup M^{2k+1}$, then

a) $$H_i(M^{2k+1}) \cong \begin{cases} H_i(N^{2k+1}), & i \neq k \\ H_k(N^{2k+1}) + (S^{\wedge} + S), & i = k \end{cases}$$

where $S = \operatorname{cok}(\mu)$ and μ is viewed as a homomorphism of free $\mathbb{Z}\pi$-modules;

b) the $(-1)^{k+1}$-linking form on $H_k(M^{2k+1})$ restricted to $S^{\wedge} + S$ is the hyperbolic form, $(S^{\wedge} + S, \varphi_h, \psi_h)$.

<u>Proof</u>. By (4B.46), there is a homeomorphism

4B.61 $$(W^{2k+2}, N^{2k+1}) \approx ((n\bar{\mathcal{C}}^{\prime\,k+1}_{\mu} \times D^k) \cup N \times I,\ N \times \{0\})$$

where the union is taken along $n\mathcal{C}'^{k}_{\mu} \times D^k \equiv n\bar{F}(\mathcal{C}^k_{\mu}) \times D^k$ and $\bar{\mathcal{C}}'^{k+1}_{\mu}$ is constructed according to μ. By construction a neighborhood U of $\bar{F}(\mathcal{C}^k_{\mu})$ is removed for surgery on $\bar{F}(\mathcal{C}^k_{\mu})$. U intersects $\bar{F}(\bar{\mathcal{C}}^{k+1}_{\mu})$ in a copy of $\mathcal{C}_{\mu} \times I$ and since, in this case, $\bar{\mathcal{C}}^{k+1}_{\mu} - (\mathcal{C}^k_{\mu} \times I) \approx \bar{\mathcal{C}}^{k+1}_{\mu}$, $\bar{F}$ furnishes a framed immersion $\bar{G}: \bar{\mathcal{C}}^{k+1}_{\mu} \to M^{2k+1}$. By (4B.61) there is a natural inclusion $\bar{\mathcal{C}}'^{k+1}_{\mu} \to W$; let $\bar{H}: \bar{\mathcal{C}}'^{k+1}_{\mu} \to M^{2k+1}$ denote this inclusion pushed out to the boundary of $n\bar{\mathcal{C}}'^{k+1}_{\mu} \times D^k$ by a vector in the complementary framing (the D^k-factor). If $\bar{H}$ and $\bar{G}$ are chosen carefully, we may assume $\bar{H}|\mathcal{C}'^{k}_{\mu} = \bar{G}|\mathcal{C}^k_{\mu}$ with compatible framings and hence obtain a framed immersion

4B.62 $$J: \mathcal{C}^{k+1}_{\mu} \to M \ , \quad \mathcal{C}^{k+1} = \bar{\mathcal{C}}^{k+1}_{\mu} \cup \bar{\mathcal{C}}'^{k+1}_{\mu}, \quad J = \bar{G} \cup \bar{H}.$$

Here $\bar{\mathcal{C}}^{k+1}_{\mu} \cup \bar{\mathcal{C}}'^{k+1}_{\mu} \approx \mathcal{C}^{k+1}_{\mu}$ because the union is taken along the "boundaries" $\mathcal{C}^k_{\mu}$ and $\mathcal{C}'^{k}_{\mu}$.

For the rest of the proof we omit superscripts from W^{2k+2}, M^{2k+1}, and N^{2k+1}. By (4B.61) and excision the map induced by the natural inclusion

4B.63 $$\ell_*: H_*(\bar{\mathcal{C}}'^{k+1}_{\mu}, \mathcal{C}'^{k}_{\mu}) \longrightarrow H_*(W,N)$$

is an isomorphism. By duality, and (4B.25),

4B.64 $$H_i(W,M) \cong \begin{cases} S^\wedge, & i = k+1 \\ 0\ , & i \neq k+1\ . \end{cases}$$

Therefore the non-trivial section of the homology sequence of the pair (W,M) is

4B.65 $$H_{k+1}(M) \rightarrowtail H_{k+1}(W) \longrightarrow H_{k+1}(W,M) \xrightarrow{\partial} H_k(M) \twoheadrightarrow H_k(W).$$

We claim ∂ is a split injection so that

4B.66 $$H_i(M) \cong \begin{cases} H_{k+1}(W,M) + H_k(W), & i = k \\ H_i(W), & i \neq k\ . \end{cases}$$

Consider the commutative diagram

$$\begin{array}{ccc} H_{k+1}(W,M) & \xrightarrow{\partial} & \tau H_k(M) \\ \downarrow \cong & & \downarrow \cong \\ H_k(W,N)^\wedge & \xrightarrow{i^\wedge} & (\tau H_k(M))^\wedge \end{array}$$

where the verticals are duality isomorphisms, $\tau H_k(M)$ denotes torsion part (2.2), and $i\colon \tau H_k(M) \to H_k(W,N)$ is induced by the inclusion $(M,\emptyset) \to (W,N)$. From this is suffices to show i is a split surjection. This follows from the commutativity of

4B.67
$$\begin{array}{ccccc} H_k(\bar{\mathcal{C}}^{k+1}_\mu \cup \bar{\mathcal{C}}'^{k+1}_\mu) & \xrightarrow{\cong} & H_k(\bar{\mathcal{C}}^{k+1}_\mu \cup \bar{\mathcal{C}}'^{k+1}_\mu, \bar{\mathcal{C}}^{k+1}_\mu) & \xleftarrow{\cong} & H_k(\bar{\mathcal{C}}'^{k+1}_\mu, \mathcal{C}'^{k}_\mu) \\ \downarrow J_* & & & & \cong \downarrow \ell_* \\ tH_k(M) & & \xrightarrow{\quad i \quad} & & H_k(W,N) \end{array}$$

where the top horizontals are isomorphisms because $H_i(\bar{\mathcal{C}}^{k+1}_\mu) = 0$, $i \geq 2$, and by excision, respectively.

By an argument similar to that above (diagram (4B.67) with W replacing M), $H_k(W) \to H_k(W,N)$ is a split surjection. So from the homology sequence of (W,N) we obtain

4B.68
$$H_i(W) \cong \begin{cases} H_k(N) + H_k(W,N), & i = k \\ H_i(N), & i \neq k. \end{cases}$$

Part a) in (4B.60) now follows from (4B.68),(4B.66) and (4B.64) with $S \cong H_k(W,N)$ and $S^\wedge \cong H_{k+1}(W,M)$.

To prove part b) observe first that $S = \operatorname{im} J_*$ by (4B.67) and $S^\wedge = \operatorname{im} \partial$ by (4B.65) and (4B.64). The assumption $\bar{F}(\dot{\bar{\mathcal{C}}}^{k+1}_\mu) \cap \bar{F}(b\bar{\mathcal{C}}^{k+1}_\mu) = \emptyset$ implies that $J(\dot{\mathcal{C}}^{k+1}_\mu) \cap J(b\mathcal{C}^{k+1}_\mu) = \emptyset$. By (4B.33) S is annihilated by the linking and self-linking forms. (Since J is framed, S supports a self-linking form). To treat $S^\wedge$ we need some notation. Let $\bar{U} = n\bar{\mathcal{C}}'^{k+1}_\mu \times D^k$, $U = n\mathcal{C}'^{k}_\mu \times D^k \equiv nF(\mathcal{C}^k_\mu) \times D^k$, and $V = \partial\bar{U} - \mathring{U}$. Let

$\bar{I}$: $(\bar{\mathcal{C}}^{k+2}_{\bar{\mu}}, \mathcal{C}^{k+1}_{\bar{\mu}}) \to (U, \partial U)$ denote the inclusion of the dual conglomerate pair in (4B.31), $\bar{I}'$: $(\bar{\mathcal{C}}^{k+2}_{\bar{\mu}}, \mathcal{C}^{k+1}_{\bar{\mu}}) \to (W,M)$ is its composition with the inclusion $(U,\partial U) \to (W,M)$ and I': $\mathcal{C}^{k+1}_{\bar{\mu}} \to M$ is the restriction of $\bar{I}'$ to $\mathcal{C}^{k+1}_{\bar{\mu}}$. We have the commutative diagram

(4B.69)
$$\begin{array}{ccccccc}
H_k(\mathcal{C}^{k+1}_{\bar{\mu}}) & \xleftarrow{\cong} & H_{k+1}(\bar{\mathcal{C}}^{k+2}_{\bar{\mu}}, \mathcal{C}^{k+1}_{\bar{\mu}}) & \underset{\cong}{\xrightarrow{\bar{I}_*}} & H_{k+1}(U,\partial U) & \underset{\cong}{\xrightarrow{\alpha}} & H_{k-1}(U)^{\wedge} \\
\downarrow{\scriptstyle I'_*} & & \downarrow{\scriptstyle \bar{I}'_*} & & \downarrow{\scriptstyle u} & & {\scriptstyle\cong}\downarrow{\scriptstyle \partial^{\wedge}} \\
H_k(M) & \xleftarrow{\partial} & H_{k+1}(W,M) & \underset{\cong}{\xleftarrow{\text{exc}}} & H_{k+1}(\bar{U}, V) & \underset{\cong}{\xrightarrow{\beta}} & H_k(\bar{U},U)^{\wedge}
\end{array}$$

where α and β are duality isomorphism, $\bar{I}_*$ is an isomorphism by (4B.43), ∂: $H_k(\bar{U},U) \to H_{k-1}(U)$ is an isomorphism since $H_i(\bar{U}) = 0$, $i \geq 2$, u is induced by the inclusion $(U,\partial U) \hookrightarrow (\bar{U}, V)$, and $(\bar{U}, V)$ (W,M) is an excision. Hence $S^{\wedge} = \mathrm{im}(I'_*)$. Once again we may apply (4B.33) (since I' is a framed imbedding) to conclude $S^{\wedge}$ is annihilated by linking and self-linking forms. This completes the proof.

<u>Chapter 4C</u>: <u>Composition of Moore Spaces</u>

Given a matrix $\mu \in M_n(\mathbb{Z}\pi, \mathbb{Q}\pi)^{\times}$ there is $t \in \mathbb{Z}$ and $\nu \in M_n(\mathbb{Z}\pi, \mathbb{Q}\pi)^{\times}$ such that $\mu\nu = tI_n$, where I_n is the $(n \times n)$-identity matrix and $M_n(\mathbb{Z}\pi, \mathbb{Q}\pi)^{\times}$ is defined in (4B.10). Viewing μ and ν as homomorphisms of free modules, there is consequently a short exact sequence ([Ma, p. 51, Ex. 6] and (2.42),

4C.1
$$\begin{array}{ccccc}
E(\nu, \mu) = \mathrm{cok}\ \nu & \rightarrowtail & \mathrm{cok}\ \mu\nu & \twoheadrightarrow & \mathrm{cok}\ \mu \\
 & & \updownarrow{\scriptstyle\cong} & & \\
 & & ((\mathbb{Z}/t\mathbb{Z})[\pi])^n & &
\end{array}$$

In a category $\mathcal{P}$ whose objects are $\mathbb{Z}\pi$-projectives, stabilization should mean addition of $\mathbb{Z}\pi$-free modules; however in $\mathfrak{T}^1_F$, a torsion analogue of $\mathcal{P}$, more general extensions must be used. Indeed, we must view the extension of cok μ by cok ν in (4C.1) as stabilization, since in general there is no way to enlarge cok μ to something simple like $((\mathbb{Z}/t\mathbb{Z})[\pi])^n$, except by non-split extensions. (An extension (4C.1) of $\mathbb{Z}\pi$-projectives would always split.)

It turns out that stabilization of a formation (2.36(c)), the analogue of stabilization of a unitary matrix, requires that we be able to pass geometrically between the "pieces", cok ν and cok μ, and their extension, cok $\mu\nu$, in (4C.1). From Chapter 4B we know that cok $\mu\nu$ is represented by $\mathcal{C}^k_{\mu\nu}$; to start Chapter 4C we define the composition $\mathcal{C}^k_\mu\mathcal{C}^k_\nu$ of $\mathcal{C}^k_\mu$ and $\mathcal{C}^k_\nu$, which is supposed to represent $E(\nu, \mu)$ (see (7.36)). The main result is (4C.19) which says that the framed imbeddings of $\mathcal{C}^k_\mu\mathcal{C}^k_\nu$ and of $\mathcal{C}_{\mu\nu}$ in a piecewise-linear manifold are in one-to-one correspondence up to isotopy. Actually we must prove somewhat more for the application in Chapter Seven: that certain homology data (4C.14) remain fixed in transition between $\mathcal{C}^k_\mu\mathcal{C}^k_\nu$ and $\mathcal{C}^k_{\mu\nu}$.

4C.2 <u>Definition</u>. Let μ and ν be $(n\times n)$-matrices in $M_n(\mathbb{Z}\pi,\mathbb{Q}\pi)^\times$ and let $\tilde{\mathcal{C}}^k_\mu$ and $\tilde{\mathcal{C}}^k_\nu$ be the corresponding conglomerates. In the usual notation ((4B.1) and (4B.11)) let $\tilde{\mathcal{C}}^k_\mu = \bigcup_{g,i} \mathfrak{M}(g,i)$ and remove from $\mathfrak{M}(g,i)$ the interior $\mathring{E}^k_{g,i}$ of a k-disc $E^k_{g,i}$, where $E^k_{gh,i} = E^k_{g,i}\cdot h$, for all $g,h \in \pi$, and $i = 1,\ldots,n$. Let

$$\{T^{k-1}_{g,i} \mid T^{k-1}_{g,i} \text{ is a } (k-1)\text{-sphere, } g \in \pi,\ i = 1,\ldots,n\}$$

be the set of bocksteins of $\tilde{\mathcal{C}}^k_\nu$ and identify $T^{k-1}_{g,i}$ with $\partial E^k_{g,i}$, for each g and i, in orientation-reversing fashion. The resulting space, $\tilde{\mathcal{C}}^k_\mu\tilde{\mathcal{C}}^k_\nu$, is the <u>composition of $\tilde{\mathcal{C}}^k_\nu$ with $\tilde{\mathcal{C}}^k_\mu$</u>. The identifications made are consistent with the π-action, so π acts freely on $\tilde{\mathcal{C}}^k_\mu\tilde{\mathcal{C}}^k_\nu$; let $\mathcal{C}^k_\mu\mathcal{C}^k_\nu$ denote the orbit space.

4C.3 <u>Proposition</u>. With the above notation

$$H_i(\tilde{\mathcal{C}}^k_\mu \tilde{\mathcal{C}}^k_\nu) \cong \begin{cases} \operatorname{cok}(\mu\nu), & i = k-1 \\ 0, & i \neq k-1, 1, 0, \end{cases}$$

where $\operatorname{cok}(\mu\nu)$ denotes the cokernel of the map of free $\mathbb{Z}\pi$-modules with matrix $\mu\nu$.

Proof. We omit all superscript k's during the proof. Let $\tilde{\mathcal{C}}_\nu = \bigcup_{g,i} \mathcal{M}(g,i)$ (where the $\mathcal{M}(g,i)$ are generalized Moore spaces) and remove from each $\mathcal{M}(g,i)$ the interior $\mathring{D}_{g,i}$ of a k-disc such that $D_{gh,i} = D_{g,i}\cdot h$, for all $g,h \in \pi$, $i = 1,\ldots,n$. Set $\tilde{D} = \bigcup_{g,i} D_{h,i}$, $\partial\tilde{D} = \bigcup_{g,i} \partial D_{g,i}$, $\tilde{E} = \bigcup_{g,i} E_{g,i}$ (see (4C.2)), and $\partial\tilde{E} = \bigcup_{g,i} \partial E_{g,i}$. From the Mayer-Vietoris sequence of the pair $((\tilde{\mathcal{C}}_\mu - \mathring{\tilde{E}}) \cup \tilde{D}, \tilde{\mathcal{C}}_\nu - \mathring{\tilde{D}})$ and (4B.4) (iii) we find $H_i(\tilde{\mathcal{C}}_\mu \tilde{\mathcal{C}}_\nu)$ as required, except possibly for $i = k, k-1$. For this, the remaining relevant part of the Mayer-Vietoris sequence is

4C.4 Diagram.

$$H_k(\tilde{\mathcal{C}}_\mu \tilde{\mathcal{C}}_\nu) \rightarrowtail H_{k-1}(\partial\tilde{E}\cup\partial\tilde{D}) \xrightarrow{m} H_{k-1}((\tilde{\mathcal{C}}_\mu - \mathring{\tilde{E}})\cup D) + H_{k-1}(\tilde{\mathcal{C}}_\nu - \mathring{\tilde{D}}) \twoheadrightarrow H_{k-1}(\tilde{\mathcal{C}}_\mu \tilde{\mathcal{C}}_\nu)$$

$H_{k-1}(\partial\tilde{E}\cup\partial\tilde{D})$ is $\mathbb{Z}\pi$-free with (ordered) basis $A = \{[\partial E_{e,1}],\ldots,[\partial E_{e,n}],[\partial D_{e,1}],\ldots,[\partial D_{e,n}]\}$. Let $b\tilde{\mathcal{C}}_\mu = \bigcup_{g,i} S_{g,i}$ (as in (4B.1)). By (4B.4)(iii) $H_{k-1}((\tilde{\mathcal{C}}_\mu - \mathring{\tilde{E}})\cup\tilde{D}) + H_{k-1}(\tilde{\mathcal{C}}_\nu - \mathring{\tilde{D}})$ is $\mathbb{Z}\pi$-free with (ordered) basis $B = \{[S_{e,1}],\ldots,[S_{e,n}],[T_{e,1}],\ldots,[T_{e,n}]\}$. Using the map μ' in Diagram (4B.5), we see that $[\partial E_{e,i}]$ is homologous in $H_{k-1}(\tilde{\mathcal{C}}_\mu - \mathring{\tilde{E}}) = H_{k-1}((\tilde{\mathcal{C}}_\mu - \mathring{\tilde{E}}) \cup \tilde{D})$ to $\sum_j [S_{e,j}]\mu_{ji}$, and that $[\partial D_{e,i}]$ is homologous in $H_{k-1}(\tilde{\mathcal{C}}_\nu - \mathring{\tilde{D}})$ to $\Sigma[T_{e,j}]\nu_{ji}$. From this it is easy to see that the $(2n\times 2n)$-matrix of m with respect to the bases A and B is

4C.5
$$\begin{pmatrix} \mu & 0_n \\ -I_n & \nu \end{pmatrix}.$$

This matrix is clearly $\mathbb{Q}\pi$-invertible, so $H_k(\tilde{\mathcal{C}}_\mu \tilde{\mathcal{C}}_\nu) = 0$ in (4C.4). Further, the matrix equation

$$\begin{bmatrix} I_n & \mu \\ 0_n & I_n \end{bmatrix} \begin{bmatrix} \mu & 0_n \\ -I_n & \nu \end{bmatrix} \begin{bmatrix} I_n & \nu \\ 0_n & I_n \end{bmatrix} = \begin{bmatrix} 0_n & \mu\nu \\ -I_n & 0_n \end{bmatrix}$$

shows that $H_{k-1}(\tilde{\mathcal{C}}_\mu \tilde{\mathcal{C}}_\nu) \equiv \operatorname{cok}(m) \cong \operatorname{cok}(\mu\nu)$.

4C.6. To define the distinguished and complementary fields on $\tilde{\mathcal{C}}^k_\mu \tilde{\mathcal{C}}^k_\nu$, we require that (in the notation of (4C.2)) the collar direction at each bockstein component $T^{k-1}_{g,i}$ agree with the collaring of $\partial E^k_{g,i}$ in $\mathfrak{M}(g,i) - \dot{E}^k_{g,i}$, and that the distinguished and complementary fields on $\tilde{\mathcal{C}}^k_\nu$, restricted to each $T^{k-1}_{g,i}$, agree with those on $\mathfrak{M}(g,i)$, restricted to $\partial E^k_{g,i}$. These fields on $\tilde{\mathcal{C}}^k_\mu \tilde{\mathcal{C}}^k_\nu$ define the corresponding fields on the quotient $\mathcal{C}^k_\mu \mathcal{C}^k_\nu$ (as in (4B.18)), and thus also the trivial normal bundle $n(\mathcal{C}^k_\mu \mathcal{C}^k_\nu) \times D^m$.

We now present a method for converting $\mathcal{C}^k_\mu \mathcal{C}^k_\nu$ to $\mathcal{C}^k_{\mu\nu}$ and vice-versa. This is the geometric tool needed to study stabilization (2.36) of a formation in Chapter Seven. Roughly speaking, the conversion of $\mathcal{C}^k_\mu \mathcal{C}^k_\nu$ to $\mathcal{C}^k_{\mu\nu}$ is accomplished by sliding $b\mathcal{C}^k_\nu$ down along $\mathcal{C}^k_\mu$ to $b\mathcal{C}^k_\mu$. The process is complicated by the inevitable introduction of "extraneous" sheets at $b(\mathcal{C}^k_{\mu\nu})$ and by the need to keep track of the homology both of the boundary of $n(\mathcal{C}^k_\mu \mathcal{C}^k_\nu) \times D^{k-1}$ and of its complement $N^{2k} - (n(\mathcal{C}^k_\mu \mathcal{C}^k_\nu) \times D^{k-1})$, for some framed imbedding $\mathcal{C}^k_\mu \mathcal{C}^k_\nu \to N^{2k}$, where N^{2k} is a p.l. manifold. For notational and conceptual simplicity we begin with constructions for Moore spaces and then generalize them to conglomerates (as in Chapter 4A and Chapter 4B).

4C.7 Proposition. Let s and t be positive integers. There is an imbedding

$$J\colon n\mathfrak{M}^k_{ts} \longrightarrow n(\mathfrak{M}^k_t \mathfrak{M}^k_s)$$

and disjoint imbedded 2-discs $\Gamma_1, \ldots, \Gamma_r \subseteq n(\mathfrak{M}_t \mathfrak{M}_s)$, $r = (s-1)(t-1)$ such that

(i) $\Gamma_i \cap J(\mathfrak{M}^k_{ts}) = \partial\Gamma_i$, $i = 1,\ldots,r$,

(ii) The distinguished field on $\mathfrak{M}_{ts}$ in $n\mathfrak{M}_{ts}$ agrees (under J) with the collar direction of $\partial\Gamma_i$ in Γ_i, $i = 1,\ldots,r$,

(iii) $\{\partial\Gamma_1,\ldots,\partial\Gamma_r\}$ represents a subset of a free generating set for $\pi_1(J\mathfrak{M}^k_{ts}) = \pi_1(\mathfrak{M}^k_{ts})$,

(iv) The inclusion $J(\mathfrak{M}_{ts}) \cup (\bigcup_i \Gamma_i) \hookrightarrow n(\mathfrak{M}_t\mathfrak{M}_s)$ is a simple homotopy equivalence.

<u>Proof</u>. For the proof we largely omit superscript k's. Let $N = \mathfrak{M}_s - R$, where N is the closure of a neighborhood of $b\mathfrak{M}_s$ in $\mathfrak{M}_s$, let $\mathfrak{M}'_t \subset n\mathfrak{M}_t$ denote $\mathfrak{M}_t$ pushed apart at its bockstein (4A.17), and form $\mathfrak{M}'_t\mathfrak{M}_s$ in the obvious way.

We work with a model for $\mathfrak{M}'_t\mathfrak{M}_s$. To construct this, let I^k denote the k-fold product of $I = [0,1]$, let $a_j = (j-1)/(2t-1) \in I$, $j = 1,\ldots,2t$, and let $D_i = I^{k-1} \times [a_{2i-1}, a_{2i}] \subseteq I^{k-1} \times I$ be k-discs in I^k, $i = 1,\ldots,t$. Let $B_i \subset \mathring{D}_i$ be a k-disc with $D_i - \mathring{B}_i \approx \partial D_i \times [0,1]$. Finally, let $K(s)$ denote the cone on s points and let p be the cone point.

Let $Y = (I^k - \bigcup_{i=1}^{t} \mathring{B}_i) \cup (K(s) \times \partial(I^k))$ where the union is taken over $\partial(I^k) \subset I^k - \cup \mathring{B}_i$ and $\{p\} \times \partial(I^k) \subset K(s) \times \partial(I^k)$.

4C.8 <u>Figure</u>.

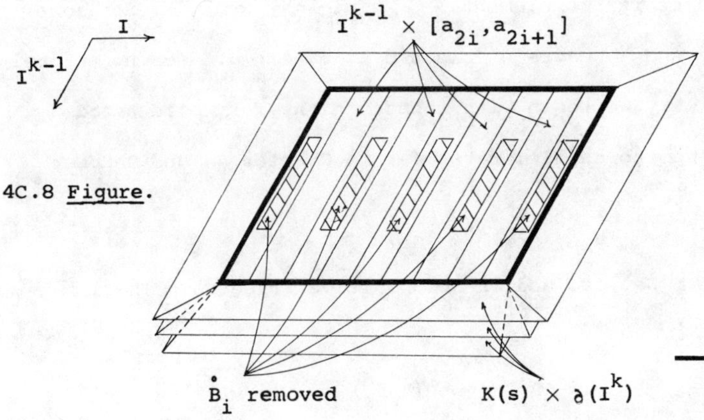

$= Y$, $t = 5$, $s = 3$, $k = 2$

▬▬: singularity of order s.

Thus there is a homeomorphism

4C.9 $$h\colon Y \xrightarrow{\approx} (\mathfrak{M}'_t\mathfrak{M}_s) - R.$$

Let $S(s)$ be the suspension of the space of s points, and let $(I^{k-1} \times [a_{2i},a_{2i+1}]) \cup \partial I^{k-1} \times K(s)$ be $I^{k-1} \times [a_{2i},a_{2i+1}]$ extended slight-slightly in the I^{k-1} direction across the "bockstein of Y", $\partial(I^k)$, where $i = 1,\ldots,t-1$. Now replace $(I^{k-1} \times [a_{2i},a_{2i+1}]) \cup \partial(I^{k-1}) \times K(s)$ by $I^{k-1} \times S_i(s)$ by replacing the $[a_{2i},a_{2i+1}]$ factor by the $S_i(s)$ factor, $S_i(s) \approx S(s)$, $i = 1,\ldots,t-1$.

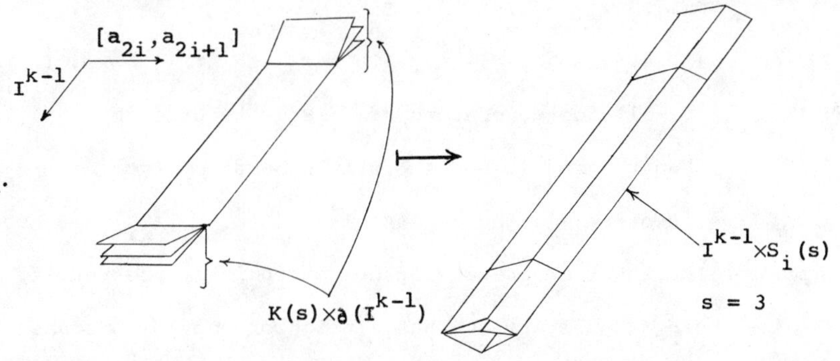

4C.10 Figure.

This can be done in nY, where nY is given by the pull back under h of the distinguished field on $\mathfrak{M}'_t\mathfrak{M}_s$. By this replacement process, Y is replaced by Y', shown in Figure 4C.11:

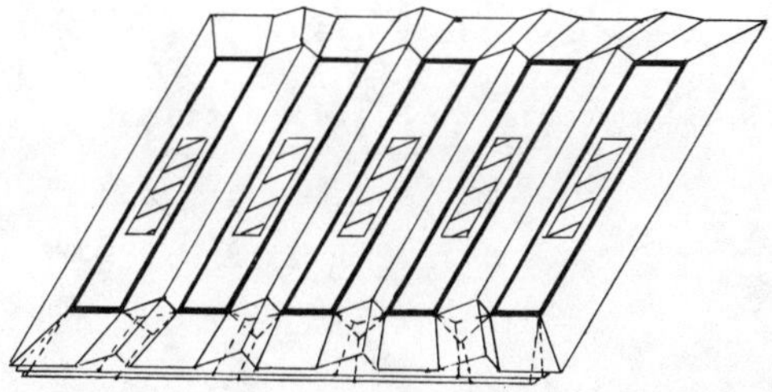

4C.11 Figure.

Y' where $\begin{cases} t = 5 \\ s = 3 \\ k = 2 \end{cases}$ ——— : singularity of order s.

The singularity of order s at points of $\partial(I^k)$ in Y has been replaced by a singularity of order s at points of $\partial(I^{k-1} \times [a_{2i-1},a_{2i}])$ in Y', $i = 1,\ldots,t$. Using h in (4C.9), we may translate this operation to one on $\mathfrak{M}_t'\mathfrak{M}_s$, taking place in $n(\mathfrak{M}_t\mathfrak{M}_s)$ (recall $\mathfrak{M}_t'\mathfrak{M}_s \subset n(\mathfrak{M}_t\mathfrak{M}_s)$). Let $\mathfrak{M}_{t,s}'$ denote the result, and let $h': Y' \to \mathfrak{M}_{t,s}'$ denote the imbedding given by h and the conversion of Y to Y'. $\mathfrak{M}_{t,s}'$ has the following two properties.

4C.12.

(i) It is homeomorphic to the generalized Moore space $\mathfrak{M}^k_{\{s,\ldots,s\}}$, whose bockstein has t components. This is clear by construction. In our model the bockstein is $\bigcup_{1\leq i\leq t} \partial(I^{k-1} \times [a_{2i-1}, a_{2i}])$ and the collar is $\bigcup_i (I^{k-1} \times [a_{2i-1}, a_{2i}] - \mathring{B}_i)$.

(ii) If we attach 2-discs Γ_j in $n\mathfrak{M}_{t,s}'$ to $h'(I^{k-1} \times S_i(s))$ ($\subseteq \mathfrak{M}_{t,s}'$) to kill the generators of $\pi_1(S_i(s))$ for each $i = 1,\ldots,t-1$, then the result has the simple homotopy type of $\mathfrak{M}_t'\mathfrak{M}_s$.

If we identify all the collars $I^{k-1} \times [a_{2i-1}, a_{2i}] - \mathring{B}_i$ to one another by orientation-preserving homeomorphisms, the result, $\mathfrak{M}_{t,s}$, is a space whose singularity set has a neighborhood homeomorphic to $S^{k-1} \times K(t,s)$ where $K(t,s)$ is $K(t)$ together with a copy of $K(s)$ attached to each of the t free ends of $K(t)$ (cf. left side of Fig. 4C.13)). By (i) above, if we slide each $K(s)$-term in the $K(t,s)$-factor of $S^{k-1} \times K(t,s)$ over to the cone point of $K(t)$, getting a neighborhood of the singularity set homeomorphic to $S^{k-1} \times K(ts)$,

4C.13 Figure. 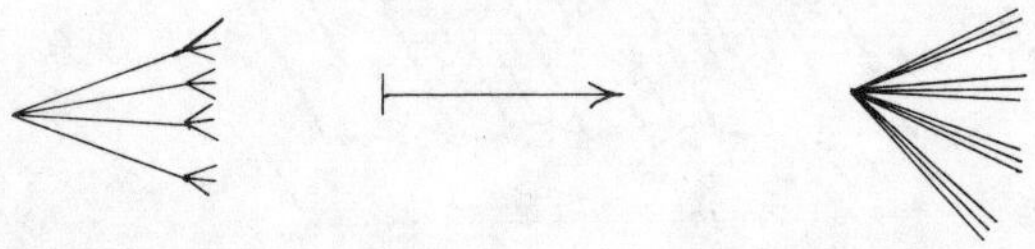 $t = 4$, $s = 3$

then $\mathfrak{M}_{s,t}$ is converted to $\mathfrak{M}_{st}$, because this will be the space obtained by identifying all t components of $b\mathfrak{M}^k_{\{s,\ldots,s\}}$ to a single $(k-1)$-sphere by orientation-preserving homeomorphisms. We have $\mathfrak{M}_{ts} \simeq \mathfrak{M}_{t,s}$ and by (ii) above $\mathfrak{M}_{t,s} \cup (\bigcup_{1\leq j\leq r} (\Gamma_j)) \simeq \mathfrak{M}_t\mathfrak{M}_s$, where the Γ_j are attached along free generators of $\pi_1\mathfrak{M}_{t,s} = \pi_1\mathfrak{M}_{ts}$ as required in (4C.7)(iii)-(iv); (4C.7) (i)-(ii) are clear by construction. All constructions (e.g., adding discs, replacing Y by Y') take place in $n(\mathfrak{M}_t\mathfrak{M}_s)$, so we get the imbedding J as required.

We now enrich this proposition. Let V be a regular neighborhood of a subspace X of the p.l. manifold N^{2k}. Consider the following data, denoted $(X,V,N^{2k})_*$, consisting of an exact sequence and a homomorphism of $\mathbb{Z}\pi$-modules:

4C.14 $$(X,V,N^{2k})_* = \begin{cases} H_k(V,\partial V) \xrightarrow{i} H_{k-1}(\partial V) \xrightarrow{j} H_{k-1}(V) \\ H_k(N - \mathring{V},\partial V) \xrightarrow{\Delta} H_{k-1}(\partial V). \end{cases}$$

4C.15 <u>Proposition</u>. Assume $k \geq 4$. Let $F\colon \mathfrak{M}^k_{ts} \to N^{2k}$ be a framed imbedding and let U_{ts} be the regular neighborhood of its image. Then there is $V_{ts} \subseteq N$, $V_{ts} = U_{ts} \cup$ (2-handles), and a framed imbedding $G\colon \mathfrak{M}^k_t\mathfrak{M}^k_s \to N^{2k}$ unique up to isotopy where V_{ts} is the regular neighborhood of $G(\mathfrak{M}^k_t\mathfrak{M}^k_s)$ and $(F\mathfrak{M}^k_{ts},U_{ts},N^{2k})_* = (G(\mathfrak{M}^k_t\mathfrak{M}^k_s),V_{ts},N^{2k})_*$. Conversely, given a framed imbedding $G\colon \mathfrak{M}^k_t\mathfrak{M}^k_s \to N^{2k}$ and its regular neighborhood $V_{ts} \subseteq N^{2k}$, there is a framed imbedding $F\colon \mathfrak{M}^k_{ts} \to V_{ts} \subseteq N^{2k}$ unique up to isotopy, whose regular neighborhood U_{ts} satisfies $(G(\mathfrak{M}^k_t\mathfrak{M}^k_s),V_{ts},N^{2k})_* = (F\mathfrak{M}^k_{ts},U_{ts},N^{2k})_*$. The process of conversion of G to F and F to G are inverse to one another up to isotopy.

<u>Proof</u>: Given $F\colon \mathfrak{M}^k_{ts} \to N^{2k}$, the circles $\partial\Gamma_1,\ldots,\partial\Gamma_r$ constructed in the model Y' (see Fig. (4C.11) and (4C.7)(ii)) bound imbedded discs $\Gamma_1,\ldots,\Gamma_r$ in N^{2k} $(r = (s-1)(t-1))$, since $\pi_1\mathfrak{M}^k_{ts} \to \pi_1 N^{2k}$ is trivial (4B.16). Since $k \geq 4$, the Γ_i are unique up to isotopy fixing $\partial\Gamma_i$. We require (imitating (4C.7)(ii)) that the collar of $\partial\Gamma_i$ in Γ_i be given by the distinguished framing, ϵ, of $\mathfrak{M}^k_{ts}$ in N. Then the $\partial\Gamma_i$, pushed out to ∂U_{ts} by ϵ, furnish framed imbedded circles $\partial\Gamma_i'$ on ∂U_{ts} with framings extending over the Γ_i. Ambient surgery on the $\partial\Gamma_i'$ in ∂U_{ts} with core 2-discs Γ_i clearly yields $U_{ts} \cup$ (2-handles), a regular neighborhood V_{ts} for $\mathfrak{M}^k_{ts} \cup (\bigcup_i \Gamma_i)$. Since the collar direction of $\partial\Gamma_i$ in Γ_i has been chosen to coincide with ϵ, and $U_{ts} \approx n\mathfrak{M}^k_{ts} \times D^{k-1}$, (4C.7)(iv) and the uniqueness

of regular neighborhoods provides $G\colon \mathfrak{M}_t^k\mathfrak{M}_s^k \to N^{2k}$ with regular neighborhood V_{ts}. Since to obtain V_{ts}, 2-handles were added to U_{ts} in N^{2k} along a subset of a free generating set for $\pi_1 U_{ts}$ and since $k \geq 4$, $(F\mathfrak{M}_{ts}^k, U_{ts}, N^{2k})_* = (G(\mathfrak{M}_t^k\mathfrak{M}_s^k), V_{ts}, N^{2k})_*$.

The converse procedure is easier because (4C.7) shows we may imbed $(n\mathfrak{M}_{ts}^k) \times D^{k-1}$ in $(n\mathfrak{M}_t^k\mathfrak{M}_s^k) \times D^{k-1}$. Again (4C.7) and the above handle-and-homology argument shows there is no change in the data (4C.14). Since the choice of Γ_i is fixed up to isotopy, by our use of the models Y and Y' in (4C.7), and the requirement $k \geq 4$, the constructions of F from G and G from F are inverse up to isotopy.

The construction of F from G in (4C.15) effectively increases the number of sheets abutting to $b\mathfrak{M}_t^k$. If $\mathfrak{M}_s^k$ in $\mathfrak{M}_t^k\mathfrak{M}_s^k$ is chosen canonically enough, the next proposition shows that F can also be constructed by "increasing the order of the bockstein" (4A.35).

4C.16 <u>Proposition</u>. Let $H\colon \mathfrak{M}_t^k \to N^{2k}$ be a framed imbedding, let $F_1\colon \mathfrak{M}_{ts}^k \to N^{2k}$ be obtained from H by increasing the order of the bockstein, and let U_1 be the regular neighborhood of $F_1(\mathfrak{M}_{ts}^k)$. Let the framed imbedding $G_2\colon \mathfrak{M}_t^k\mathfrak{M}_s^k \to N^{2k}$ be obtained from H by requiring that $G_2|\mathfrak{M}_s^k$ be a trivial imbedding (4B.59), let $F_2\colon \mathfrak{M}_{ts}^k \to N^{2k}$ be obtained from G_2 by (4C.15) and let U_2 be the regular neighborhood of $F_2(\mathfrak{M}_{ts}^k)$. Then $(F_1\mathfrak{M}_{ts}^k, U_1, N^{2k})_* = (F_2\mathfrak{M}_{ts}^k, U_2, N^{2k})_*$.

<u>Proof</u>. It suffices by (4C.15) to show that if $G_1\colon \mathfrak{M}_t^k\mathfrak{M}_s^k \to N^{2k}$ is the imbedding produced from F_1 by (4C.15), then G_1 is isotopic to G_2. We will keep the notation used in the proofs of (4C.15) and (4A.32). Let $\tilde{H}_i\colon \mathfrak{M}_t^k \to N$, $i = 1,\ldots,s-1$ be the $(s-1)$ copies of H used in (4A.32) (denoted F there) and let $\tilde{\mathfrak{e}}_i$ be the fields used to create the $\tilde{H}_i$. ($\tilde{\mathfrak{e}}_i = \frac{i|\tilde{\mathfrak{e}}|}{s-1}\tilde{\mathfrak{e}}$ in the notation of (4A.32)). Recall that $F_1(\mathfrak{M}_{ts}^k) = H(\mathfrak{M}_t^k) \mathbin{\hat{\#}} \tilde{H}_1(\mathfrak{M}_t^k) \mathbin{\hat{\#}} \cdots \mathbin{\hat{\#}} \tilde{H}_{s-1}(\mathfrak{M}_t^k)$, where "$\hat{\#}$" means that connected

sum is taken along the $\widetilde{H}_i(\dot{\mathfrak{M}}_t^k)$ and $H(\dot{\mathfrak{M}}_t^k)$, and that the bocksteins $H(b\mathfrak{M}_t^k)$, $\widetilde{H}_i(b\mathfrak{M}_t^k)$ are all identified to a single $(k-1)$-sphere in orientation-preserving fashion.

Push the $\widetilde{H}_i$ back from the bockstein along $H(\mathfrak{M}_t^k)$ (reversing the arrow in (4C.13)) to get $\mathfrak{M}_{t,s} \to N$ (compare Fig. (4A.34)).

4C.17 Figure.

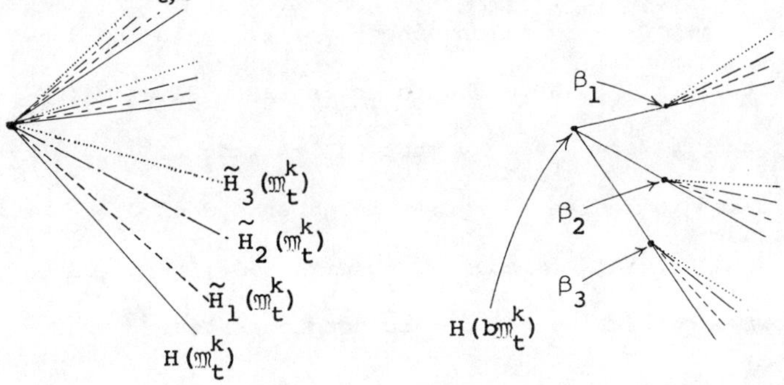

Let $\beta_1,\ldots,\beta_t$ be the $(k-1)$-spheres forming the new singularity set away from $H(b\mathfrak{M}_t^k)$. Further modify the $\tilde{\epsilon}_i$ so that they are continuous and have length zero on $\cup\beta_i$ and on the side of $H(\mathfrak{M}_t^k) - (\cup\beta_i)$ containing $H(b\mathfrak{M}_t^k)$. Now observe that the copies of $I^{k-1} \times S_j(s)$ in $\mathfrak{M}_{t,s}$ (cf. the model Y' in (4C.11)) can be chosen so that, if $S_j(s) = \ell_{1,j} \cup\cdots\cup \ell_{s,j}$ where the $\ell_{i,j}$ are line segments appropriately identified at their end points ($j=1,\ldots,t-1$), then $\ell_{1,j}$ connects β_j to β_{j+1} in $H(\dot{\mathfrak{M}}_t^k)$ and $\ell_{i+1,j} = \tilde{\epsilon}_i(\ell_{i,j})$, $i = 1,\ldots,s-1$. Thus to convert $I^{k-1} \times S_j(s)$ to $I^{k-1} \times [a_{2i}, a_{2i+1}] \cup \partial(I^{k-1}) \times K(s)$ (reversing the arrow in Fig. (4C.10)), we can take a regular neighborhood of $\bigcup_{i,j} \ell_{i,j}$ and collapse this regular neighborhood down to the neighborhood of $\bigcup_j \ell_{1,j} \subseteq H(\mathfrak{M}_t^k)$ along the $\tilde{\epsilon}_i$-lines running through it. By construction, the resultant space is the image of $G_1\colon \mathfrak{M}_t^k\mathfrak{M}_s^k \to N^{2k}$.

It is clear by construction that

4C.18
$$G_1|\mathfrak{M}_t^k - \dot{E}_1 = G_2|\mathfrak{M}_t^k - \dot{E}_2$$

where E_1 and E_2 are the k-discs removed from $\mathfrak{M}_t^k$ to form the domain $\mathfrak{M}_t^k\mathfrak{M}_s^k$ for G_1 and G_2. Let $B \subseteq H(\dot{\mathfrak{M}}_t^k)$ be a small k-disc removed to connect

$H(\mathfrak{M}_t^k)$ to $\tilde{H}_1(\mathfrak{M}_t^k)$ in the construction of F_1. Then $H(\mathfrak{M}_t^k) - \mathring{B} \subseteq F_1(\mathfrak{M}_{ts}^k)$ and, by our construction of G_1, $H(\mathfrak{M}_t^k) - \mathring{B} \subseteq G_1(\mathfrak{M}_t^k\mathfrak{M}_s^k)$ and $\mathring{B} \cap G_1(\mathfrak{M}_t^k\mathfrak{M}_s^k) = \emptyset$. From $H(\mathfrak{M}_t^k) - \mathring{B} \simeq H(b\mathfrak{M}_t^k) \vee$ (bouquet of circles) $\simeq G_1(\mathfrak{M}_t^k\mathfrak{M}_s^k - \mathfrak{M}_s^k) = G_1(\mathfrak{M}_t^k - \mathring{E}_1)$, it follows that $(H(\mathfrak{M}_t^k) - \mathring{B}) - G_1(\mathfrak{M}_t^k - \mathring{E}_1) \approx (\partial B) \times I$. Let $\bar{B}$ denote the k-disc $H(\mathfrak{M}_t^k) - G_1(\mathfrak{M}_t^k - \mathring{E}_1)$. Then $G_1(\mathfrak{M}_s^k) \cup$ (s copies of $\bar{B}$) is the image of a map $J\colon S^k \to N$ (obtained by filling in the s k-discs removed from S^k to form $\mathfrak{M}_s^k$) where the union is taken along $G_1(b\mathfrak{M}_s^k) = \partial\bar{B}$. It is left to the reader to verify that J extends to $\bar{J}\colon D^{k+1} \to N$. ($\bar{J}(D^{k+1})$ may be taken to lie in a regular neighborhood of $G_1(\mathfrak{M}_t^k\mathfrak{M}_s^k)$.) This means $G_1|\mathfrak{M}_s^k$ is a "trivial" imbedding, so with (4C.15) the proof is complete.

Now we generalize (4C.15) to conglomerates.

4C.19 <u>Proposition</u>. Let μ and $\nu \in M_n(\mathbb{Z}\pi,\mathbb{Q}\pi)^\times$. Let $F\colon \mathcal{C}_{\mu\nu}^k \to N^{2k}$ be a framed imbedding and let $U_{\mu\nu}$ be the regular neighborhood of its image. Then F can be converted to a framed imbedding $G\colon \mathcal{C}_\mu^k\mathcal{C}_\nu^k \to N^{2k}$, unique up to isotopy, so that if $V_{\mu,\nu}$ denotes the regular neighborhood of its image, then $(F\mathcal{C}_{\mu\nu}^k, U_{\mu\nu}, N^{2k})_* = (G(\mathcal{C}_\mu^k\mathcal{C}_\nu^k), V_{\mu,\nu}, N^{2k})_*$. Conversely, given a framed imbedding $G\colon \mathcal{C}_\mu^k\mathcal{C}_\nu^k \to N^{2k}$, it can be converted to a framed imbedding $F\colon \mathcal{C}_{\mu\nu}^k \to N^{2k}$, unique up to isotopy, without altering the data (4C.14). The processes converting F to G and G to F are inverse to one another, up to isotopy.

<u>Proof</u>. The proof is carried out in progressively more general steps. Let $\mathcal{C}_\mu^k = \bigcup_{i=1}^n \mathfrak{M}(i)$, $\tilde{\mathcal{C}}_\mu^k = \bigcup_{g,i} \mathfrak{M}(g,i)$, $\mathcal{C}_\nu^k = \bigcup_i \mathfrak{N}(i)$, $\tilde{\mathcal{C}}_\nu^k = \bigcup_{g,i} \mathfrak{N}(g,i)$, $b\mathcal{C}_\mu^k = \bigcup S_i^{k-1}$, $b\tilde{\mathcal{C}}_\mu^k = \bigcup_{g,i} S_{g,i}^{k-1}$, $b\mathcal{C}_\nu^k = \bigcup T_i^{k-1}$ (see (4C.2) for this notation). For the proof we omit explicit mention of the imbeddings F and G and work instead with their images, denoted $\mathcal{C}_{\mu\nu}^k$ and $\mathcal{C}_\mu^k\mathcal{C}_\nu^k$, respectively.

<u>Step 1</u>: $n = 1$, $\mu = \mu_{11} = tg$, $\nu = \nu_{11} = sh$, where $s,t \in \mathbb{Z}$, $g,h \in \pi$. Thus $\mathcal{C}_\mu^k \approx \mathfrak{M}_{|t|}^k$ and $\mathcal{C}_\nu^k \approx \mathfrak{M}_{|s|}^k$. This is essentially Prop. (4C.15).

Step 2(a): $n = 1$, $\mu = \mu_{11} = tg$, $\nu = \nu_{11} = \Sigma\, m_h h$. In this case $\mathcal{C}_\mu^k \approx \mathfrak{M}_{|t|}^k$ and $\mathcal{C}_\nu^k = \mathfrak{N}(1)$, an identified Moore space. Recall that $\mathfrak{N}(1) = (\bigcup_g \mathfrak{N}(g,1))/\pi$ (4B.13), where $\mathfrak{N}(e,1) \approx \mathfrak{M}^k_{\{|m_h|\,|h\in\pi\}}$. Thus $\mathfrak{N}(1)$ is a punctured k-sphere with boundary components all identified to a single (k-1)-sphere by homeomorphisms which are not necessarily orientation-preserving:

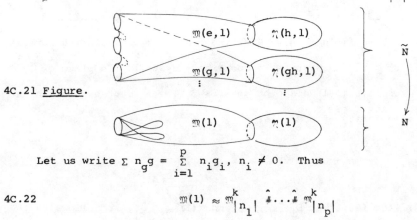

4C.20 Figure.

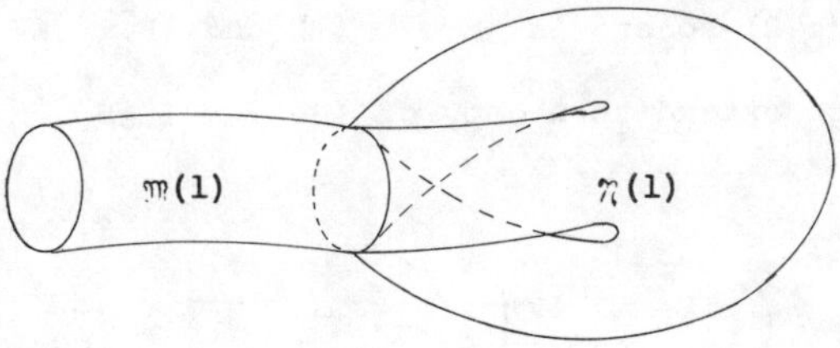

The proof of (4C.15) clearly applies here without change to yield the result. $\mathfrak{N}(1)$ may carry non-trivial elements of $\pi_1(N^{2k})$ but it is easy to verify that no 2-handle in (4C.15) need be attached to such elements coming from $\pi_1(\mathfrak{M}(1)\mathfrak{N}(1))$.

Step 2(b): $n = 1$, $\mu = \mu_{11} = \Sigma\, n_g g$, $\nu = \nu_{11} = sh$. This case is also very much like (4C.15), although the fundamental construction ($Y \to Y'$ in the proof of (4C.15)) must be applied twice so we give the details. We have $\mathcal{C}_\mu^k = \mathfrak{M}(1)$, an identified Moore space and $\mathcal{C}_\nu^k = \mathfrak{N}(1) \approx \mathfrak{M}_{|s|}^k$

4C.21 Figure.

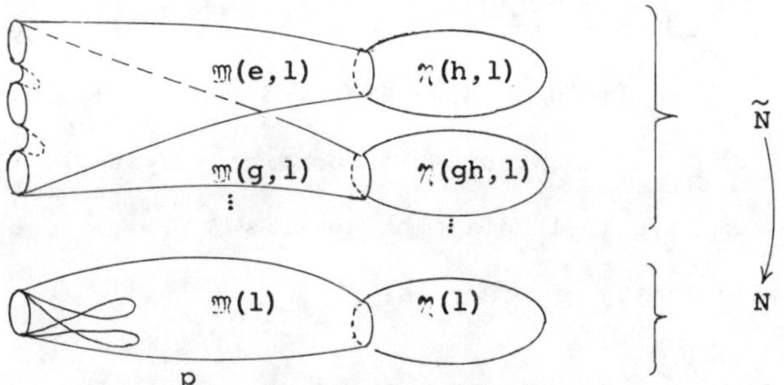

Let us write $\Sigma\, n_g g = \sum_{i=1}^{p} n_i g_i$, $n_i \neq 0$. Thus

4C.22 $$\mathfrak{M}(1) \approx \mathfrak{M}_{|n_1|}^k \,\hat{\#}\cdots\hat{\#}\, \mathfrak{M}_{|n_p|}^k$$

where by "$\hat{\#}$" we mean that connected sum is taken along the interiors and that the bocksteins on the right are all identified to $b\mathfrak{M}(1) \approx S^{k-1}$ by homeomorphisms (not necessarily orientation-preserving). Let

$\Sigma_1,\ldots,\Sigma_{p-1} \subseteq \mathcal{M}(1) \subseteq N^{2k}$ be the (k-1)-spheres along which the connected sum (4C.22) is taken. We may assume $\Sigma_i \cap E^k$ is a (k-1)-disc, where E^k is the k-disc removed from $\mathcal{M}(1)$ to glue in $\mathcal{N}(1)$. Set

4C.23 $$B_i^{k-1} = \Sigma_i - (\Sigma_i \cap \mathring{E}), \quad i = 1,\ldots,p-1.$$

∂B_i^{k-1} is a (k-2)-sphere in $\partial E \equiv b\mathcal{N}(1)$ and $(B_i^{k-1}, \partial B_i^{k-1}) \subseteq (\mathcal{M}(1), b\mathcal{N}(1))$ may be assumed to extend to a copy of $(B_i^{k-1} \times I, \partial B_i^{k-1} \times I)$ in $(\mathcal{M}(1), b\mathcal{N}(1))$.

4C.24 Figure.

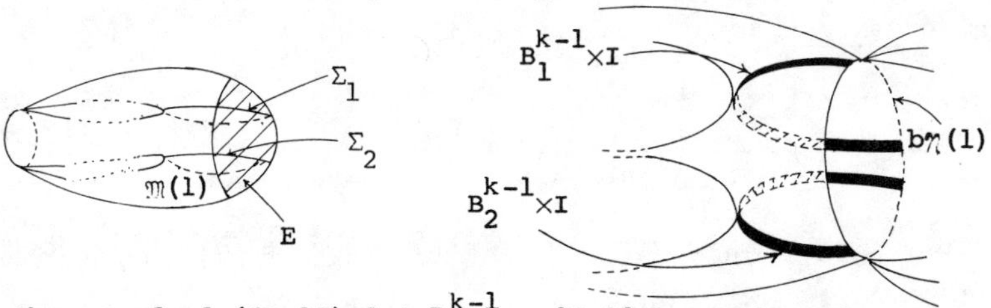

Following the proof of (4C.15) let $B_i^{k-1} \times [0,1]$ here play the role of $I^{k-1} \times [a_{2i}, a_{2i+1}]$ there. Thus if $Z_i = B_i^{k-1} \times [0,1] \underset{\partial B_i^{k-1}\times[0,1]}{\cup} \partial B_i \times [0,1] \times K(s)$ represents $B_i^{k-1} \times [0,1]$ extended slightly across $b\mathcal{N}(1)$ into $\mathcal{N}(1)$ we replace

4C.25 $$Z_i \quad \text{by} \quad B_i^{k-1} \times S(s) \qquad \text{(see Fig. (4C.10))}.$$

As in Prop. (4C.15) this may be done so that the distinguished field on Z_i is replaced by a field on $B_i \times S(s)$ induced by upward pointing normals on $S(s)$ in $S(s) \subseteq R^2$ (in a suitable coordinate system). In place of $b\mathcal{M}(1) \approx S^{k-1}$ we have (p-1) disjoint spheres $\beta_1,\ldots,\beta_{p-1}$ at each of which there is a singularity of order s:

4C.26 Figure.

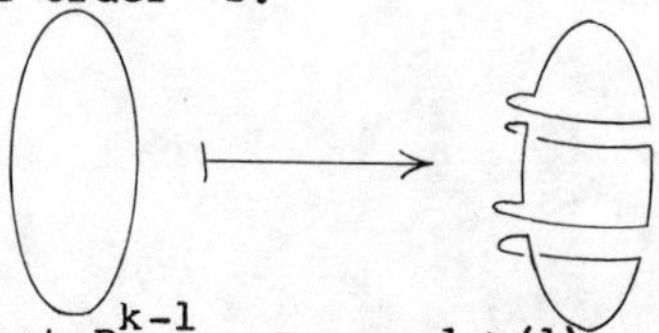

From (4C.22), the fact that $B_i^{k-1} \subseteq \Sigma_i$, and $\mathcal{N}(1) \approx \mathcal{M}^k_{|s|}$, it follows that the result of (4C.25) is homeomorphic to

4C.27 $$(\mathcal{M}^k_{|n_1|}, \mathcal{M}^k_{|s|}) \;\hat{\#}\; (\mathcal{M}^k_{|n_2|}, \mathcal{M}^k_{|s|}) \;\hat{\#}\cdots\hat{\#}\; (\mathcal{M}^k_{|n_p|}, \mathcal{M}^k_{|s|})$$

where the connected sum is taken along (k-1)-spheres $\sigma_1,\ldots,\sigma_{p-1}$ in the copies of $\mathring{\mathfrak{M}}_s^k \subseteq \mathfrak{M}^k_{|n_i|}\mathfrak{M}^k_{|s|}$.

4C.28 Figure.

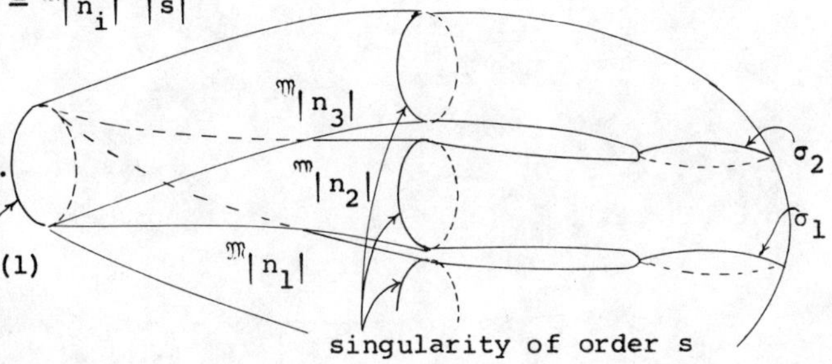

This was the first application of the method of (4C.15). Now we apply (4C.15) itself to each of the terms on the right in (4C.27). The result is clearly $\mathcal{C}^k_{\mu\nu}$ as required and the data of (4C.14) are unchanged in the transition from $\mathcal{C}^k_\mu\mathcal{C}^k_\nu$ to $\mathcal{C}^k_{\mu\nu}$ because we need only attach 2-discs to $\mathcal{C}^k_{\mu\nu}$ to recover $\mathcal{C}^k_\mu\mathcal{C}^k_\nu$ (up to simple homotopy). Details are left to the reader.

Step 3: $n = 1$, $\mu = \mu_{11} = \sum_{i=1}^{p} n_i g_i$, $\nu = \nu_{11} = \sum_{j=1}^{q} m_j h_j$. (Suppose again that no n_i or m_j is zero.) Here $\mathcal{C}^k_\nu = \mathfrak{N}(1)$ and $\mathcal{C}^k_\mu = \mathfrak{M}(1)$ are both identified Moore spaces; the proof begins with a combination of Steps 2(a) and 2(b). First, as in Step 2(b) convert $\mathfrak{M}(1)\mathfrak{N}(1)$ to $(\mathfrak{M}^k_{|n_1|}\mathfrak{N}_1(1)\hat{\#}\ldots\hat{\#}(\mathfrak{M}^k_{|n_p|}\mathfrak{N}_p(1))$ where each $\mathfrak{N}_i(1)$ is homeomorphic to $\mathfrak{N}(1)$ (see Fig. (4C.28)). Then use Step 2(a) to push the bockstein of each identified Moore space $b\mathfrak{N}_i(1)$ down along $\mathfrak{M}^k_{|n_i|}$ to $b\mathfrak{M}^k_{|n_i|}$.

Here we encounter a difficulty. Fix integers i and j, where $1 \leq i \leq p$ and $1 \leq j \leq q$, and let $h = g_i g_j$. The number of sheets abutting to $S^{k-1}_{h,1}$ in $\tilde{N}$ from $\mathfrak{M}(g_j,1)$ is $|n_i|$. After sliding $b\mathfrak{N}_i(1)$ down to $b\mathfrak{M}^k_{|n_i|}$ we change this to $|n_i||m_j|$:

4C.29 Figure.

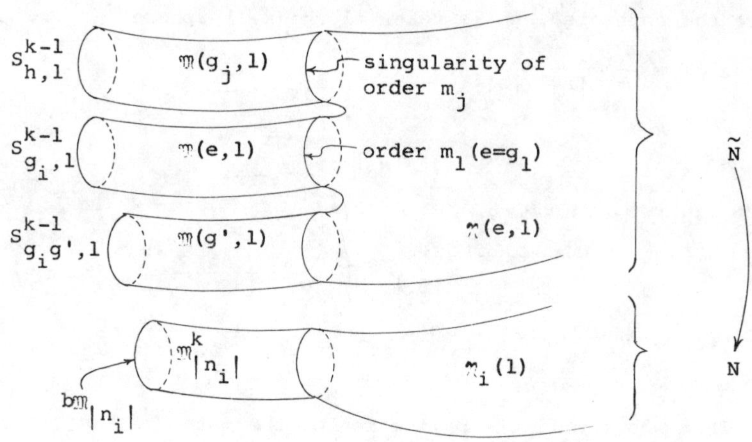

Hence letting i and j vary but fixing $h = g_ig_j$ yields altogether $\sum_{g_ig_j=h} |n_i||m_j|$ sheets abutting to S_h^{k-1} in $\tilde{N}$. Algebraically (i.e., counting orientations) the contribution is $\sum_{g_ig_j=h} n_im_j$, the coefficient of h in $\mu\nu$. Hence in $\tilde{N}$, the process gives a space $\mathfrak{M}_{\mu\nu}(e,1)$ which is a punctured sphere with not necessarily orientation preserving identifications on the boundary, but where the equation of chains

$$\partial(\mathfrak{M}_{\mu\nu}(e,1)) = \sum_g r_g\, S^{k-1}_{g,1}$$

remains valid, where r_g = coefficient of g in $\mu\nu$. Hence by construction (4B.1) the problem that remains is to reduce the number of sheets abutting to $S^{k-1}_{h,1}$ from $\sum_{g_ig_j=h} |n_im_j|$ to $|\sum_{g_ig_j=h} n_im_j|$. Once this is done we will have converted $\mathfrak{M}_{\mu\nu}(e,1)$ to the appropriate term in the union (4B.11) corresponding to $\tilde{\mathcal{C}}^k_{\mu\nu}$. Projecting this conversion to N or making it equivariant in $\tilde{N}$ completes the conversion of $\mathcal{C}^k_\mu\mathcal{C}^k_\nu$ to $\mathcal{C}^k_{\mu\nu}$.

Assume $|\Sigma\, n_im_j| \neq \Sigma|n_im_j|$, where the sums are over i,j such that g_ig_j is fixed and equals $h \in \pi$. Working in $\tilde{N}$, this means

4C.30
$$\mathfrak{M}_{\mu\nu}(e,1) = \mathfrak{M}_1 \,\hat{\#}\, \mathfrak{M}_2$$

where $\mathfrak{M}_1$ and $\mathfrak{M}_2$ are punctured spheres with identifications (not necessarily orientation-preserving) on their boundaries,

$\mathfrak{m}_2 \approx \{S^k - (\mathring{D}_1^k \cup \mathring{D}_2^k)\}/(\partial D_1^k \approx -\partial D_2^k)$ (the boundary components are identified in orientation reversing fashion), and $b\mathfrak{m}_2 = S_{h,1}^{k-1} \subseteq b\mathfrak{m}_{\mu\nu}(e,1)$. Thus $\mathfrak{m}_2 \approx S^k \times S^1$. Let

4C.31 $$B_1 \subseteq \mathring{\mathfrak{m}}_1 \,, \quad B_2 \subseteq \mathring{\mathfrak{m}}_2$$

be the k-discs removed from $\mathfrak{m}_1$ and $\mathfrak{m}_2$ to form (4C.30). Then

4C.32 $$\mathfrak{m}_2 - \mathring{B}_2 \simeq S_{h,1}^{k-1} \vee S^1.$$

Let $r\colon S^{k-1} \times C(t) \xrightarrow{\approx} W$ be a homeomorphism to a neighborhood W of $S_{h,1}^{k-1}$ in $b\mathfrak{m}_1$ and let $s\colon S^{k-1} \times R^2 \times R^{k-1} \xrightarrow{\approx} R(S_{h,1}^{k-1})$ be a homeomorphism to a regular neighborhood of $S_{h,1}^{k-1}$ in $\tilde{N}$. Since $\mathfrak{m}_1$ is framed imbedded, with respect to appropriate choices of r and s, its imbedding into $\tilde{N}$ has the form (near the bockstein)

4C.33 $$\mathrm{id} \times c\colon S^{k-1} \times C(t) \longrightarrow S^{k-1} \times R^2 \times R^{k-1}$$

where $\mathrm{id}\colon S^{k-1} \to S^{k-1}$ is the identity and $c\colon C(t) \to R^2 \times R^{k-1}$ is the imbedding of (4A.5) (furnishing the distinguished framing) followed by the standard inclusion $R^2 \to R^2 \times R^{k-1}$.

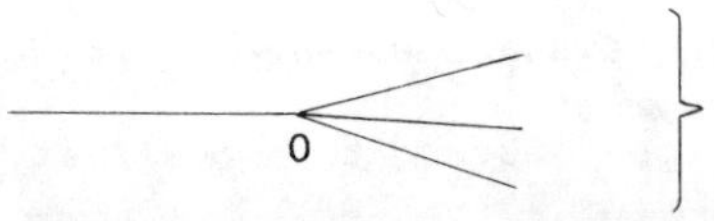

$C(t), t = 3$

Adding two more lines to this picture and joining their ends by a line segment

4C.34 <u>Figure</u>.

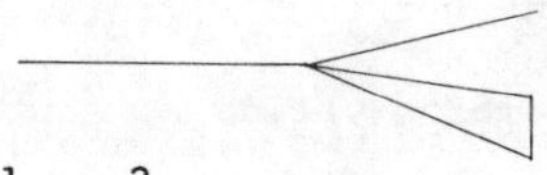

we obtain $c'\colon C(t) \vee S^1 \to R^2$ and hence

4C.35 $$\mathrm{id} \times c'\colon S^{k-1} \times (C(t) \vee S^1) \longrightarrow S^{k-1} \times R^2 \times R^{k-1}.$$

Since $S^{k-1} \times (C(t) \vee S^1) = (S^{k-1} \times C(t)) \cup (S^{k-1} \times S^1)$ where the union is taken along $S^{k-1} \times \{0\} \subseteq S^{k-1} \times C(t)$ and $S^{k-1} \times \{p\} \subseteq S^{k-1} \times S^1$ $(p \in S^1)$,

(4C.35) constructs a model for $\mathfrak{M}_2$ imbedded in a neighborhood of $S^{k-1}_{h,1}$. Since the S^1-term in (4C.35) bounds a 2-disc in $\tilde{N}$ (disjoint from $\mathfrak{M}_1$), we may assume that the given $\mathfrak{M}_2 - \dot{B}_2$ in (4C.32) coincides with the model in (4C.35). This done, we remove $\mathfrak{M}_2 - \dot{B}_2$ from (4C.30) and put back $\mathring{B}_1$ (see (4C.31)) where $\mathring{B}_2$ was in the model (4C.35)

$$s(id \times c')(S^{k-1} \times S^1) \subseteq \tilde{N}$$

for $\mathfrak{M}_2$, $s: S^{k-1} \times R^2 \times R^{k-1} \overset{\approx}{\to} R(S^{k-1}_{h,1})$.

4C.36 Figure.

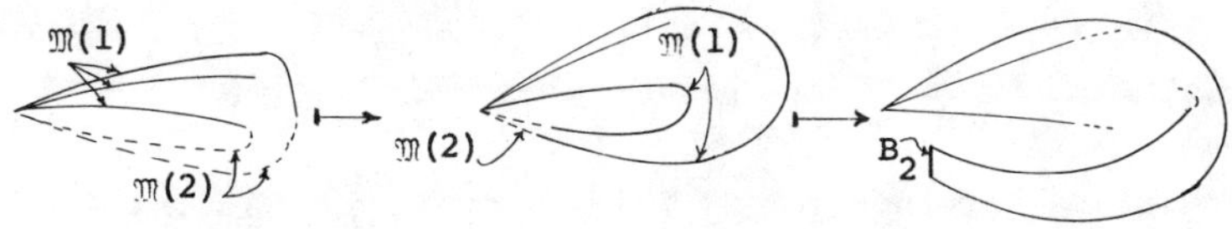

This eliminates two extraneous sheets at $S^{k-1}_{h,1}$; in addition, the specific model shows how to create a pair of extraneous sheets converting $\mathfrak{M}_1$ to $\mathfrak{M}_1 \hat{\#} \mathfrak{M}_2$: change (4C.33) to (4C.35), take out $\mathring{B}_1$ and $\mathring{B}_2$ (4C.31) and add in a tube

4C.37 $$T \approx S^{k-1} \times I$$

connecting ∂B_1 to ∂B_2. It is straightforward to verify that the processes of creating and eliminating extraneous sheets are well-defined and inverse to one another, up to isotopy. Using the models it is easy to keep track of the framings. As mentioned above, we project these operations to N to make the required changes in $\mathcal{T}(1)$ and $\mathfrak{M}(1)$. If $t = 0$ in the above discussion (i.e., if $\mathfrak{M}_1$ does not abut to $S^{k-1}_{h,1}$ in (4C.30)), no mention of $C(t)$ is made and the argument remains valid.

It remains to show that adding extraneous sheets does not change the data (4C.14); by what was done above, this also covers the case of eliminating extraneous sheets. Now we work in N. Let U denote the regular neighborhood in N of a framed imbedded identified Moore space $\mathfrak{M}$ and let U' denote the regular neighborhood of $\mathfrak{M}'$, where $\mathfrak{M}'$ is $\mathfrak{M}$ with two additional extraneous sheets, as constructed above. Then U' may be obtained from U by

a) first adding (in N) a 1-handle (the S^1-term in (4C.35));

4C.38 b) adding a k-handle (the "top" cell B_2 in $S^{k-1} \times S^1 \approx \mathfrak{M}_2$);

c) adding a 1-handle (the filled-in tube $\bar{T} \approx D^k \times I$, where T is in (4C.37) and $D^k \times \{0\} = B_1$, $D^k \times \{1\} = B_2$); and

d) subtracting a k-handle (to take out $\dot{D}^k \times I$ from $\bar{T}$; a neighborhood of $\bar{T}$ is homeomorphic to $\bar{T} \times I^{k-1} \approx D^k \times I \times I^{k-1}$ and the k-handle to be removed is $\{c\} \times I \times I^{k-1}$ extended a little at both ends of the I-factor, where c is the center of D^k. See Figure (4C.39)).

4C.39 Figure.

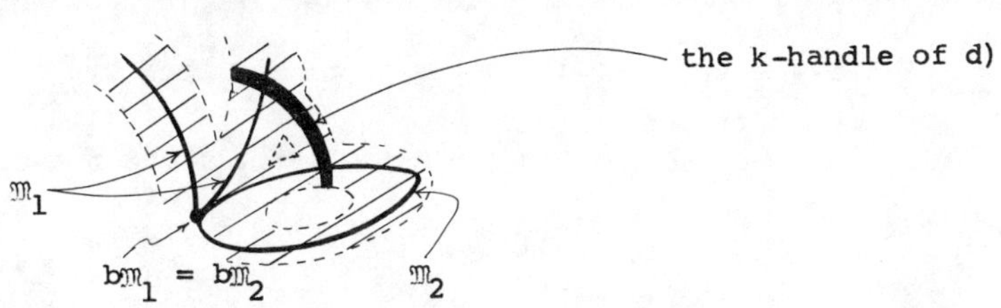

The handle addition in a) has no effect on the data of (4C.14). Let V denote U together with the handles in a) and b). Then a Mayer-Vietoris argument shows

4C.40
$$H_i(V) \cong \begin{cases} H_i(U), & i \neq k \\ H_k(U) + \mathbb{Z}\pi, & i = k. \end{cases}$$

where the $\mathbb{Z}\pi$-term is represented by $[\mathfrak{M}_2]$, $\mathfrak{M}_2 \approx S^{k-1} \times S^1$. $[\mathfrak{M}_2]$ is in the image of $H_k(\partial V) \to H_k(V)$ (push $\mathfrak{M}_2$ out to ∂V by any field from the complementary framing), and by duality applied to (4C.40),

4C.41
$$H_i(V,\partial V) \cong \begin{cases} H_i(U,\partial U), & i \neq k \\ H_k(U,\partial U) + \mathbb{Z}\pi, & i = k. \end{cases}$$

Hence the homology exact sequence of $(V,\partial V)$, (4C.40) and (4C.41) show

4C.42 $$\{H_k(V,\partial V) \rightarrowtail H_{k-1}(\partial V) \text{ — } H_{k-1}(V)\}$$
$$= \{H_k(U,\partial U) + \mathbb{Z}\pi \xrightarrow{i+Id} H_{k-1}(\partial U) + \mathbb{Z}\pi \longrightarrow H_{k-1}(U)\}.$$

It is verified in a similar way that

4C.43 $$\{H_k(N-\mathring{V},\partial V) \longrightarrow H_{k-1}(\partial V)\}$$
$$= \{H_k(N-\mathring{U},\partial U) \xrightarrow{\Delta+0} H_{k-1}(\partial U) + \mathbb{Z}\pi\}.$$

The 1-handle added in (4C.38)(c) can have no effect on the data of (4C.14). It is left to the reader to verify that the handle removal in (4C.38)(d) removes the $\mathbb{Z}\pi$-terms in (4C.40), (4C.41), (4C.42), and (4C.43). This shows that the data of (4C.14) for U agrees with that for U' as claimed. This completes the proof for Step 3.

Step 4: $n > 1$, $\mu,\nu \in M_n(\mathbb{Z}\pi,\mathbb{Q}\pi)^{\times}$. Recall the notation set at the beginning of the proof of (4C.19). Express $\mathfrak{M}(1)$ as a connected sum

4C.44 $$\mathfrak{M}(1) = \mathfrak{M}_1(1) \# \ldots \# \mathfrak{M}_n(1)$$

of identified Moore spaces where $\mathfrak{M}_i(1)$ abuts only to S_i^{k-1}, $i = 1,\ldots,n$. Let E_1 be the k-disc removed from $\mathfrak{M}(1)$ to attach T_1^{k-1}. We may assume that E_1 intersects in discs each of the (k-1)-spheres along which the connected sum (4C.44) is taken (in the manner of Step 2(b)).

4C.45 Figure.

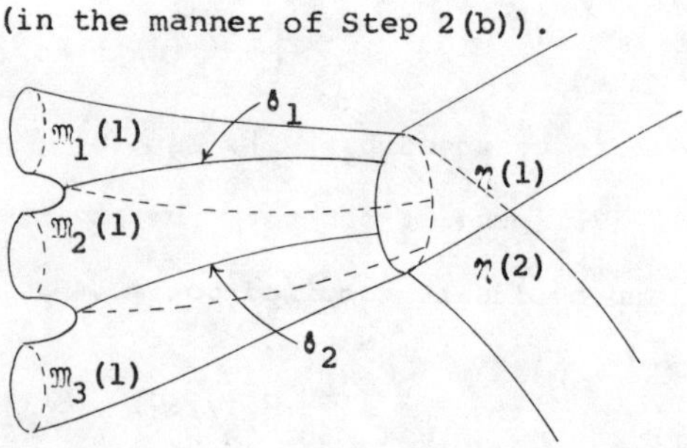

Pushing apart in neighborhoods of the complements δ_i of these intersections (see (4C.24)) leaves $\eta(1),\ldots,\eta(n)$ attached to each $\mathfrak{M}_1(i)$, $i = 1,\ldots,n$:

4C.46 <u>Figure</u>.

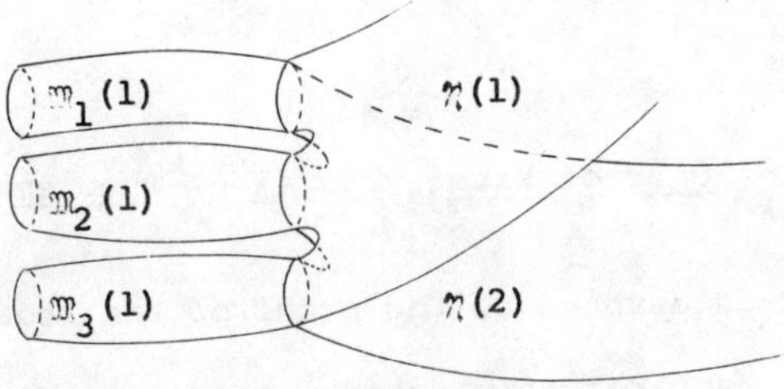

Next push the singularities at $b\eta(i)$ across $\mathfrak{M}_j(1)$ to $b\mathfrak{M}_j(1)$, $i,j = 1,\ldots,n$, as in Steps 2(a) and 3. Doing this for $\mathfrak{M}(2),\ldots,\mathfrak{M}(n)$ completes the conversion $\mathcal{C}^k_\mu \mathcal{C}^k_\nu$ to $\mathcal{C}^k_{\mu\nu}$. The requirements of (4C.19) are satisfied here because we have verified them in the previous steps. This completes the proof of (4C.19).

Chapter Five: The $(\pi-\pi)$ Theorem.

This chapter is devoted to the proof of the local version of [W2,3.3]. Our Theorem (5.3) requires more careful geometric techniques, but as the reader familiar with [W2] will recognize, if a conglomerate is thought of as a replacement for a sphere, our proof follows that of Wall. We begin with an improvement of (1.20). Recall from (1.12) the definition of "local surgery problem".

5.1 Proposition. Let $(g;b): (N,\partial N;\nu_N) \to (Y,X;\xi)$ be a local surgery problem where $\dim N = 2k+2 \geq 8$, and $\pi_1 X \xrightarrow{\cong} \pi_1 Y$. Then $(g;b)$ is locally bordant to $(h;c): (N',\partial N';\nu_{N'}) \to (Y,X;\xi)$ where the only non-zero kernels for h appear in the exact sequence

$$K_{k+1}(N',\partial N') \rightarrowtail K_k(\partial N') \twoheadrightarrow K_k(N'),$$

and $\pi_1(\partial N) \cong \pi_1 N \cong \pi_1 Y \cong \pi_1 X$ (induced by g and inclusions). If $(g;b)$ is assumed to satisfy $K_i(N) = K_i(\partial N) = 0$, $i < k$, then $\partial N'$ may be obtained from ∂N by surgery on $\bar{H}(\mathcal{C}^k_\mu)$, where $\bar{H}: \bar{\mathcal{C}}^{k+1}_\mu \to \partial N$ is a framed imbedding.

Proof. By (1.20) we may assume the non-zero kernels for g appear in the exact sequence

$$K_{k+1}(N) \rightarrowtail K_{k+1}(N,\partial N) \longrightarrow K_k(\partial N) \longrightarrow K_k(N) \twoheadrightarrow K_k(N,\partial N).$$

Since $K_k(N)$ is $\mathbb{Z}$-torsion, there is $t \in \mathbb{Z}$ and framed immersions $F_i: \mathfrak{M}^{k+1}_t \to N$ such that $\{[F_i(b\mathfrak{M}^{k+1}_t)]\}$ generates $K_k(N)$, $i = 1,\ldots,n$ (use (3.17) and 4A.21). By (3.20), there is a copy of $\bar{\mathfrak{M}}^{k+1}_t \subseteq \mathfrak{M}^{k+1}_t$ such that the $F_i|\bar{\mathfrak{M}}^{k+1}_t$ are disjoint imbeddings into a disc of N. Remove the $F_i(\dot{\bar{\mathfrak{M}}}^{k+1}_t)$ from $F_i(\mathfrak{M}^{k+1}_t)$ and replace them with disjointly imbedded copies $\mathfrak{M}^k_t \times I \subseteq N$ where $\mathfrak{M}^k_t \times \{0\}$ is attached to the boundary (4A.1) of $F_i(\bar{\mathfrak{M}}^{k+1}_t)$, and $\mathfrak{M}_t^k \times \{1\}$ is in a given $(2k-1)$-disc of ∂N and bounds a copy of $\mathfrak{M}_t^{-k+1}$ in ∂N, $i = 1,\ldots,n$. This gives framed immersions

$$G_i: (\mathfrak{M}_t^{-k+1}, \mathfrak{M}_t^k) \longrightarrow (N, \partial N)$$

where the $G_i|\mathfrak{M}_t^k$ are disjoint framed trivial imbeddings (4B.59), and the relative classes $[G_i(\mathfrak{M}_t^{-k+1})]$ generate $K_k(N,\partial N)$, since $K_k(N) \to K_k(N,\partial N)$ is surjective.

By general position, the G_i may be assumed to have at most isolated double points in $\mathring{\mathfrak{M}}_t^{k+1}$, which is a manifold. Hence, using $\pi_1(\partial N) \cong \pi_1(N)$ and "piping" ([W2,p.40]), where double points are piped off to $G_i(\mathring{\mathfrak{M}}_t^k)$ (also a manifold), we may set $\mathcal{C}_{tI_n}^{-k+1}$ equal to the disjoint union of n copies of $\mathfrak{M}_t^{-k+1}$ and obtain a framed imbedding

$$G: (\mathcal{C}_{tI_n}^{-k+1}, \mathcal{C}_{tI_n}^k) \longrightarrow (N, \partial N)$$

with $G_*: H_k(\tilde{\mathcal{C}}_{tI_n}^{k+1}, \tilde{\mathcal{C}}_{tI_n}^k) \to K_k(N,\partial N)$ surjective. Since $G(\mathcal{C}_{tI_n}^{-k+1})$ carries no elements of $\pi_1 N$, if we let U denote its regular neighborhood, we can arrange that g map $(U, U \cap \partial N)$ to $(D^{2k+2}, D^{2k+1}) \subset (Y,X)$, where $(Y - \mathring{D}^{2k+2}, X - \mathring{D}^{2k+1}) \approx (Y,X)$. Setting $N' = N - \mathrm{Int}(U)$ and $h = g|N'$ where $\mathrm{Int}(U) = \mathring{U} \cup (U \cap \partial N)$ and $\mathring{U}$ is the manifold interior of U, we obtain

5.2 $$(h;c): (N', \partial N'; \nu_N) \to (Y, X; \xi).$$

Note that $(N', \partial N') \hookrightarrow (N, U \cup \partial N)$ is an excision and that

$$\begin{aligned} K_\ell(U \cup \partial N, \partial N) &\cong K_\ell(U, U \cap \partial N) \qquad \text{(excision)} \\ &\cong K_\ell(G(\mathcal{C}_{tI_n}^{-k+1}), G(\mathcal{C}_{tI_n}^k)) \\ &\cong \begin{cases} ((\mathbb{Z}/t\mathbb{Z})[\pi])^n, & \ell = k \\ 0, & \ell \neq k \end{cases} \qquad \text{by (4B.25)(ii).} \end{aligned}$$

From these two facts, the surjectivity of G_* in dimension k, and the exact sequence of the triple $(N, U \cup \partial N, \partial N)$, it follows easily that $K_i(N', \partial N') = 0$, $i \neq k+1$. By duality, $K_i(N') = 0$, $i \neq k$, so clearly $K_i(\partial N') = 0$, $i \neq k$. It remains to show that $(g;b)$ is locally bordant to $(h;c)$. For this set $W = N \times I - (\mathrm{Int}(U) \times (\frac{1}{2},1])$, where $\mathrm{Int}(U)$ is defined above and $I = [0,1]$. Let $V = \{\partial N \times I - \overline{\mathring{U} \cap \partial N} \times (\frac{1}{2},1]\} \cup (U \times \frac{1}{2}) \cup (\partial U - (\partial U \cap \partial N))$, where ∂U denotes the manifold-boundary of U. Clearly $\partial W = N \cup V \cup N'$. Extending $g \cup h$ to $\Gamma\colon (W,N,N',V) \to (Y\times I, Y\times\{0\}, Y\times\{1\}, X\times I)$ (essentially by restriction of $g \times I$) yields a bordism between $(g;b)$ and $(h;c)$. Since $V \approx \partial N \times I \cup U$, where the union is taken along $U \cap \partial N$ in $\partial N \times \{1\}$ and in U, $\Gamma|V$ is the bordism for surgery on the conglomerate $G(\mathcal{C}^k_{tI_n})$ ((4B.53)) by (4B.46.) This verifies the second assertion in (5.1) and shows $K_*(V) \otimes \mathbb{Q}\pi \equiv 0$; as observed above Γ is essentially the restriction of $g \times I$, so $K_*(W) \otimes \mathbb{Q}\pi \equiv 0$. Hence Γ is a local bordism (1.12)(b) as required, and by (1.19)(a), $(h;c)$ is a local surgery problem.

Now we come to the main result of this chapter, the "$(\pi$-$\pi)$-Theorem."

5.3 Theorem. Let $(g;b)\colon (N, \partial_+N, \partial_-N; \nu_N) \to (Y, X_+, X_-; \xi)$ be a local surgery problem where $\partial N = \partial_+N \cup \partial_-N$, $g|(\partial_-N, \partial(\partial_-N))$ is a homotopy equivalence of pairs, $\pi_i X_+ \overset{\cong}{\to} \pi_i Y$ for $i = 0,1$, and $\dim(N) \geq 9$. Then $(g;b)$ is locally bordant relative to ∂_-N to a homotopy equivalence of triads.

Proof. As in [W2,§4] ∂_-N plays no role in the proof and, for notational simplicity, is omitted. Thus we assume given $(g;b)\colon (N, \partial N; \nu_N) \to (Y, X; \xi)$. By (1.20) and assumption we take Y and X to be connected, while g and inclusions induce isomorphisms $\pi_1\partial N \cong \pi_1 N \cong \pi_1 Y \cong \pi_1 X$. We consider the cases where the dimension of N is odd or even separately.

$\dim N = 2k+1$. By (1.20) we may assume that the non-zero kernels of g appear in the exact sequence

$$K_k(\partial N) \rightarrowtail K_k(N) \longrightarrow K_k(N,\partial N) \twoheadrightarrow K_{k-1}(\partial N).$$

By (2.8), $K_k(N,\partial N) \in \mathfrak{N}_F^1$. By (4B.26) there is $\mu \in M_n(\mathbb{Z}\pi,\mathbb{Q}\pi)^\times$ and a framed immersion $\bar{F}$: $(\bar{\mathcal{C}}_\mu^{k+1},\mathcal{C}_\mu^k) \to (N,\partial N)$ such that

$$\bar{F}_*: H_k(\tilde{\bar{\mathcal{C}}}_\mu^{k+1},\tilde{\mathcal{C}}_\mu^k) \xrightarrow{\cong} K_k(N,\partial N).$$

Set $\bar{\mathcal{C}}_\mu^{k+1} = \bigcup_{i=1}^n \bar{\mathfrak{M}}(i)$ and $\mathcal{C}_\mu^k = \bigcup_{i=1}^n \mathfrak{M}(i)$ (unions of identified Moore spaces as in (4B.14)). Let $\bar{F}_i = \bar{F}|\bar{\mathfrak{M}}(i)$: $\bar{\mathfrak{M}}(i) \to N$, $F_i = \bar{F}|\mathfrak{M}(i)$: $\mathfrak{M}(i) \to \partial N$ and $F = \bar{F}|\mathcal{C}_\mu^k$: $\mathcal{C}_\mu^k \to \partial N$. For the remainder of the proof set

$$\bar{\mathcal{C}} = \bar{\mathcal{C}}_\mu^{k+1}, \quad \mathcal{C} = \mathcal{C}_\mu^k, \quad \text{where } \dim N = 2k+1.$$

We want to convert $\bar{F}$ to a framed imbedding.

By general position we may assume F: $\mathcal{C} \to \partial N$ has at most isolated double points in $F_i(\dot{\mathfrak{M}}(i)) \cap F_j(\dot{\mathfrak{M}}(j)) \subseteq \partial N$ (when $i = j$, "$\cap$" means "self-intersection"). Again we may assume $\bar{F}_i(b\bar{\mathfrak{M}}(i)) \cap \bar{F}(b\bar{\mathfrak{M}}(j)) = \emptyset$ and $\bar{F}_i(b\bar{\mathfrak{M}}(i)) \cap \bar{F}_j(\mathring{\bar{\mathfrak{M}}}(j))$ consists of isolated double points in N, $i,j = 1,\ldots,n$. Further, using a local model (of the bockstein) for the immersion $\bar{F}_i$, it is easy to see that in a neighborhood of a point $p \in \bar{F}_i(b\bar{\mathfrak{M}}(i)) \cap \bar{F}_j(\mathring{\bar{\mathfrak{M}}}(j))$, the double point set may be taken to be homeomorphic to $C(t)$, for some $t \in \mathbb{Z}$ (see (4A.3) for $C(t)$).

5.4 <u>Figure</u>

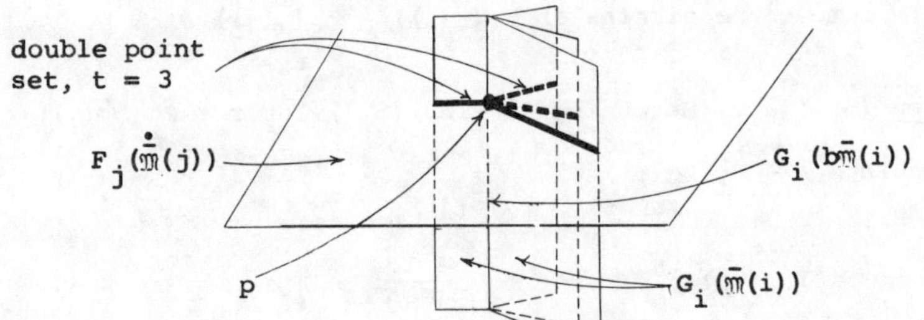

Finally, since the $\dot{\bar{\mathfrak{M}}}(i)$ are manifolds, $\bar{F}_i(\dot{\bar{\mathfrak{M}}}(i)) \cap \bar{F}_j(\dot{\bar{\mathfrak{M}}}(j))$ consists of circles and open line segments; near a double point $q \in F_i(\mathfrak{M}(i)) \cap F_j(\mathfrak{M}(j)) \subset \partial N$ the double point set $\bar{F}_i(\bar{\mathfrak{M}}(i) - b\bar{\mathfrak{M}}(i)) \cap \bar{F}_j(\bar{\mathfrak{M}}(j) - b\bar{\mathfrak{M}}(j))$ is a half-open interval in N with endpoint q. Hence Figure (5.5) presents a picture of the double point structure in $F_i(\bar{\mathfrak{m}}(i)) \cap F_j(\bar{\mathfrak{m}}(j))$.

5.5 <u>Figure</u>

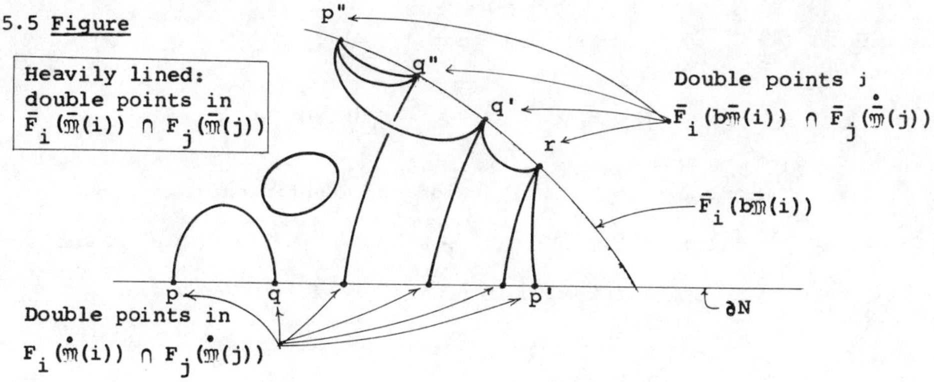

Once we have modified $\bar{F}$: $(\bar{\mathcal{C}}, \mathcal{C}) \to (N, \partial N)$ through regular homotopies to an imbedding, we will subtract a regular neighborhood of its image from $(N, \partial N)$, completing the proof of (5.3) in case dim $N = 2k + 1$, just as we did in (5.1).

To imbed $\bar{\mathcal{C}}$ there are three steps:

5.6. (a) remove the double points of $\bar{F}_i(b\bar{\mathfrak{m}}(i)) \cap \bar{F}_j(\dot{\bar{\mathfrak{m}}}(j))$ by converting them to double points in $F_i(\dot{\mathfrak{m}}(i)) \cap F(\dot{\mathfrak{m}}(j))$;

(b) remove the double points of $F_i(\dot{\mathfrak{m}}(i)) \cap F_j(\dot{\mathfrak{m}}(j))$ by converting them to circles in $\bar{F}_i(\dot{\bar{\mathfrak{m}}}(i)) \cap \bar{F}(\dot{\bar{\mathfrak{m}}}(j))$;

(c) remove the circles in $\bar{F}_j(\dot{\bar{\mathfrak{m}}}(i)) \cap \bar{F}_j(\dot{\bar{\mathfrak{m}}}(j))$.

<u>Step</u> (5.6)(a). Referring to Fig. (5.7) let $r \in \bar{F}_i(b\bar{\mathfrak{m}}(i)) \cap \bar{F}_j(\dot{\bar{\mathfrak{m}}}(j))$. Choose points $s \in F_i(b\mathfrak{m}(i))$

5.7 <u>Figure</u>

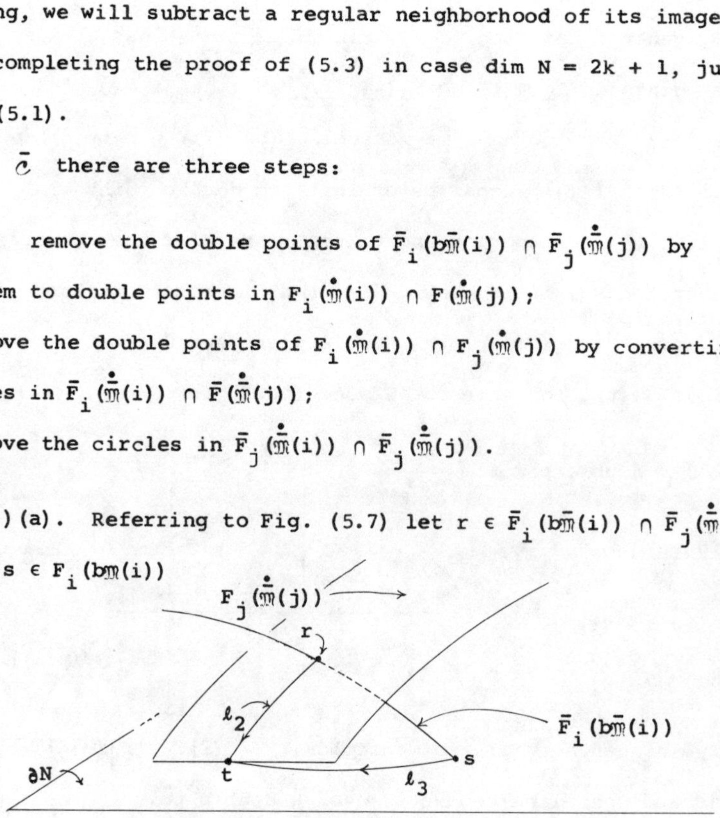

and $t \in F_j(\dot{\mathfrak{m}}(j))$, and paths ℓ_1 in $\bar{F}_i(b\bar{\mathfrak{m}}(i))$ from r to s and ℓ_2 in

$\bar{F}_j(\dot{\bar{\mathfrak{M}}}(j))$ from r to t so that ℓ_1 and ℓ_2 miss all double points except r. Since $\pi_1(\partial N) \overset{\cong}{\to} \pi_1(N)$ the path $\ell_1^{-1}\ell_2$ is homotopic relative to $\{t,s\}$ to a path ℓ_3 in ∂N connecting s to t. Since $k \geq 3$ there is an imbedded 2-disc D^2 such that $\partial D^2 = \ell_2^{-1}\ell_1\ell_3$ and $\dot{D}^2 \cap \bar{F}(\bar{\mathcal{C}}) = \emptyset$. Push a neighborhood of ℓ_2 in $\bar{F}(\bar{\mathfrak{M}}(j))$ across D^2 so that r is eliminated, without creating any new intersections. The change in the double point set near r is shown below (compare Fig. (5.5)):

5.8 Figure

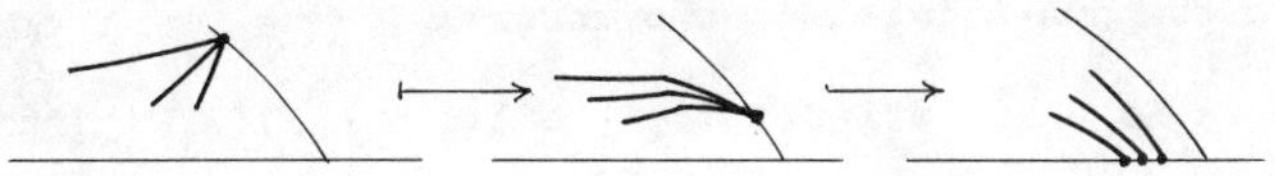

In ∂N the picutre is

5.9 Figure

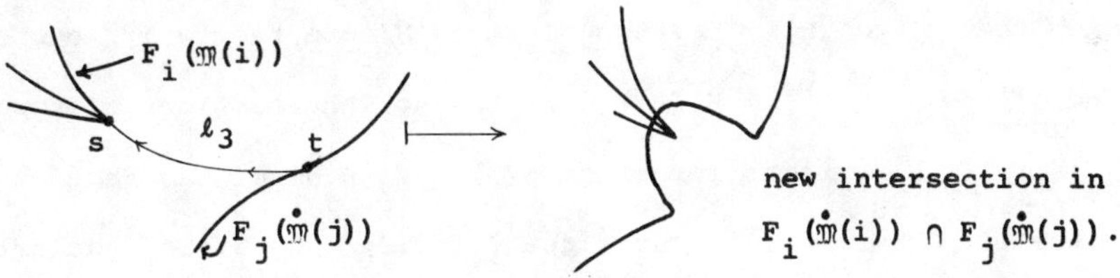

Step (5.6)(b). The previous step converts the (global) double point set (Fig. (5.5)) to a 1-manifold whose boundary consists of isolated points in $F_i(\dot{\mathfrak{M}}(i)) \cap F_j(\dot{\mathfrak{M}}(j))$ occuring in pairs $\{p,q\}$ connected by an arc $I \subseteq \bar{F}_i(\bar{\mathfrak{M}}(i)) \cap \bar{F}_j(\bar{\mathfrak{M}}(j))$, (see Fig. (5.5)). Using $\pi_1(\partial N) \overset{\cong}{\to} \pi_1(N)$, $k \geq 3$, and the well known argument in [W2,p.41], $\{p,q\}$ is removed from the double point set while converting I to a circle in $\bar{F}_i(\dot{\bar{\mathfrak{M}}}(i)) \cap \bar{F}_j(\dot{\bar{\mathfrak{M}}}(j))$.

Step (5.6)(c). It remains to eliminate circle intersections. For this we need the following lemma which classifies them. Since we have removed intersections from the bocksteins $\bar{F}_j(b\bar{\mathfrak{M}}(j))$ and "boundaries" $F_i(\mathfrak{M}(i))$ of the $\bar{F}_i(\bar{\mathfrak{M}}(i))$ we can try to carry out the remaining geometric constructions outside a neighborhood of these subsets. This can be done in our case, so the following lemma describing intersections and regular homotopies of closed manifolds will be applicable.

5.10 Lemma. Let N^n be a p.l. manifold and let $G\colon P^p \to N^n$ and

$H: Q^q \to N^n$ be immersions of closed 2-connected p.l. manifolds in transverse position, where $p + q = n + 1$ and $p,q \geq 5$.

(i) There are two types of circle intersections: (a) a circle in in $G(P^p) \cap H(Q^q)$ (resp. in the self-intersection of G, if $n = 2p - 1$) which bounds a disc both in $G(P^p)$ and $H(Q^q)$ (resp. which bounds discs in both intersecting sheets of $G(P^p)$); (b) if $n = 2p - 1$, a circle self-intersection C in $G(P^p)$ spanning $t \in \pi_1(N)$, where $t^2 = 1$ and $G^{-1}(C) \to C$ is the non-trivial double cover.

(ii) There is an invariant $\alpha(G,H) \in (\mathbb{Z}_2 \times \pi_2 N)[\pi_1 N]$ ("the $\mathbb{Z}_2 \times \pi_2 N$-group ring of $\pi_1 N$") of the regular homotopy class of G and H such that if $\alpha(G,H) = 0$, then G and H are regularly homotopic to G' and H' where $G'(P^p) \cap H'(Q^q)$ contains no intersections of type (a) above.

(iii) There is an invariant $\beta(G) \in \mathbb{Z}_2[\pi_1 N]$ of the regular homotopy class of G ($n = 2p - 1$) such that if $\beta(G) = 0$, G is regularly homotopic to G', where G' has no circle intersections of type (b) above.

<u>Proof</u>. The proof of (i) is in [W2,p.77]. Part (ii) is in [HW,VII.3] or [HQ,3.5,3.6]. We sketch the construction of α in [HW] for the reader's convenience. Having chosen base points, an element $g_C \in \pi_1 N$ is associated to each $C \subseteq G(P^p) \cap H(Q^q)$ in the usual way; $\alpha_C \in \mathbb{Z}_2 \times \pi_2 N$ is constructed as follows. Since $p,q \geq 5$ and $\pi_1 P^p = 0 = \pi_1 Q^q$ there are imbedded 2-discs $D_1 \subseteq G(P^p)$ and $D_2 \subseteq H(Q^q)$ such that $\partial D_1 = C = \partial D_2$. Let R_1 and R_2 be regular neighborhoods of D_1 and D_2 in $G(P^p)$ and $H(Q^q)$, respectively. Then $R_1 \cap R_2 = C$ and the normal bundle $\nu(C \subseteq G(P^p))$ to C in $G(P^p)$ may be framed in two ways:

(*): $\nu(C \subseteq R_1)$ is framed since R_1 is a disc

(**): $\nu(H(Q^q) \subseteq N^n)|_C$ is trivialized

by the inclusion $C \subseteq R_2 \subseteq H(Q^q)$ since R_2 is a disc. These two framings differ by an element of $\pi_1(SO(p-1) = \mathbb{Z}_2$, which is by definition the $\mathbb{Z}_2$-component of α_C; the 2-sphere $D_1 \cup D_2$ defines the π_2N-component of α_C which is well defined since $\pi_2 P^p = 0 = \pi_2 Q^q$. We set $\alpha(G,H) = \Sigma\, \alpha_C g_C$ where the summation is over type (a) circles C. Part (iii) in (5.1) is implicit in [Sh 2,§2]; $\beta(G)$ simply counts mod 2 the number of circle self-intersections of G spanning $t \in \pi_1(N)$, for each t such that $t^2 = 1$.

5.11 Remark. (a) Suppose $\pi_2 N = 0$. Spanning a 3-disc D^3 across $D_1 \cup D_2$ one might hope to imitate the Whitney device by pulling the intersection apart across D^3; the $\mathbb{Z}_2$-component of α_C is the obstruction to doing this. (b) In case $n = 2p - 1$ in (5.10) (so $p = q$), it is useful to interpret $\alpha(G,H)$ as an element of $\Omega_1^{fr}(\Omega(N,*))$, the framed bordism group of the loop space of N. This is in fact how $\alpha(G,H)$ is defined in [HQ,3.6]. Later a certain one-dimensional double point set will occur which is null-bordant; it will follow that its α-invariant is zero.

5.12 Definition. The circle intersections in (5.10)(i) will be called type (a) and type (b) respectively.

According to [Sh 2, Lemma 8.2], there is a regular homotopy $G_i\colon \mathfrak{M}(i) \times I \to \partial N \times I$ which fixes a neighborhood of $b\mathfrak{M}(i)$, moves only 1-dimensional subsets of $\mathring{\mathfrak{M}}(i)$ and satisfies $G_i|\mathfrak{M}(i) \times \{0\} = F_i$, $G_i|\mathfrak{M}(i) \times \{1\}$ is a framed imbedding, and $\beta(G_i) = \beta(\bar{F}_i)$. By appending $\partial N \times I$ to N along $\partial N \times \{0\}$ and $G_i(\mathfrak{M}(i) \times I)$ to $\bar{F}_i(\bar{\mathfrak{M}}(i))$ we obtain

$$\bar{H}_i\colon (\bar{\mathfrak{M}}(i), \mathfrak{M}(i)) \longrightarrow (N, \partial N)$$

with $\beta(\bar{H}_i) = 0$. Doing this for each $i = 1,\ldots,n$, we may apply (5.10)(iii) to remove all type (b) intersections. Similarly, the realization result for type (a) intersections [HW,p.227] permits the cancellation of type (a) intersections. Thus $\bar{F}$ has been converted to a framed imbedding $\bar{H}: (\bar{\mathcal{C}},\mathcal{C}) \to (N,\partial N)$. As remarked above, the proof of (5.3) in case $\dim N = 2k + 1$ is completed exactly as is that of (5.1) by subtracting a regular neighborhood of $\bar{H}(\bar{\mathcal{C}})$ from N.

<u>dim N = 2k</u>. By (5.1) we may assume that the non-zero kernels of $(N^{2k},\partial N^{2k})$ appear in the short exact sequence

$$K_k(N^{2k},\partial N^{2k}) \rightarrowtail K_{k-1}(\partial N^{2k}) \twoheadrightarrow K_{k-1}(N^{2k}).$$

By (2.8) $K_k(N^{2k},\partial N^{2k}) \in \mathfrak{D}^1_F$ and by (4B.26) there is $\mu \in M_n(\mathbb{Z}\pi,\mathbb{Q}\pi)^\times$, and a framed immersion $\bar{F}: (\bar{\mathcal{C}}^{k+1}_\mu,\mathcal{C}^k_\mu) \to (N^{2k},\partial N^{2k})$ such that

5.13 $$\bar{F}_*: H_k(\tilde{\bar{\mathcal{C}}}^{k+1}_\mu,\tilde{\mathcal{C}}^k_\mu) \xrightarrow{\cong} K_k(N^{2k},\partial N^{2k}).$$

Let $\bar{\mathcal{C}}^{k+1}_\mu = \bigcup_{i=1}^n \bar{\mathfrak{M}}(i)$, $\mathcal{C}^k_\mu = \bigcup_{i=1}^n \mathfrak{M}(i)$ and assume $\bar{F}$ is in transverse position. This time our goal is only to make $\bar{F}|\mathcal{C}^k_\mu: \mathcal{C}^k_\mu \to \partial N^{2k}$ an imbedding (up to regular homotopy). To simplify notation set

$$\bar{\mathcal{C}} = \bar{\mathcal{C}}^{k+1}_\mu,\ \mathcal{C} = \mathcal{C}^k_\mu\ ;\ F = \bar{F}|\mathcal{C};$$

$$N = N^{2k},\ \partial N = \partial N^{2k};\ \bar{F}_i = \bar{F}|\bar{\mathfrak{M}}(i),\ F_i = \bar{F}|\mathfrak{M}(i).$$

In general position, the intersections and self-intersections of the $\bar{F}_i|b\bar{\mathfrak{M}}(i)$ consist of isolated points and may be removed by "piping" [W2,p40]. The double point set $\bar{F}_i(b\bar{\mathfrak{M}}(i)) \cap \bar{F}_j(\bar{\mathfrak{M}}(j) - b\bar{\mathfrak{M}}(j))$ is a 1-manifold (the transverse intersection of manifolds-with-boundary) with boundary in $F_i(b\mathfrak{M}(i)) \cap F_j(\dot{\mathfrak{M}}(j))$; hence points in the latter intersection occur in pairs connected by an arc in the former, and thus may be removed by the technique of (5.6)(b). Thus the double point set $\bar{F}_i(b\bar{\mathfrak{M}}(i)) \cap \bar{F}_j(\bar{\mathfrak{M}}(j) - b\bar{\mathfrak{M}}(j))$ may be assumed to consist of type (a) circles, for any i and j, by (5.10). Using the realization theorem [HW,p.227] (as in the case $\dim N = 2k + 1$ treated above), we may append to ∂N a copy of $\partial N \times I$ and, in the latter, the track of a regular homotopy of $F_j = \bar{F}_j|\mathfrak{M}(j)$, not moving $b\mathfrak{M}(j)$ and introducing enough type (a) intersections to cancel those in $\bar{F}_i(b\bar{\mathfrak{M}}(i) \cap \bar{F}_j(\bar{\mathfrak{M}}(j) - b\bar{\mathfrak{M}}(j))$. Thus we assume that the double points of the immersion $\bar{F}: \bar{\mathcal{C}} \to N$ occur in $\bar{F}(\bar{\mathcal{C}} - b\bar{\mathcal{C}})$. Since $\bar{\mathcal{C}} - b\bar{\mathcal{C}}$ is a manifold with boundary $\mathcal{C} - b\mathcal{C}$, this double point set is a 2-manifold, with boundary circles in $F(\mathcal{C} - b\mathcal{C}) \subseteq \partial N$.

The removal of type (b) circles from $F(\mathcal{C} - b\mathcal{C})$ is accomplished in two lemmas.

5.14 Lemma. If C_1 and C_2 are two type (b) circles in the boundary of a connected component E of the self-intersection of some fixed $\bar{F}_i: \bar{\mathfrak{M}}(i) \to N$, then they define the same $g \in \pi_1 N$ (with respect to the counting process of (5.10)(iii)).

Proof. Choose base points $p \in C_1$, $q \in C_2$ and let p_m and q_m, $m = 1,2$, be the inverse images of p and q in $\mathfrak{M}(i)$ under F_i. Let $\tilde{\ell}_1$ be a path from p_1 to p_2 in $F_i^{-1}(C_1)$ and $\tilde{\ell}_3$, a path in $F_i^{-1}(C_2)$ from q_2 to q_1. Let $\tilde{\ell}_2$ (resp. $\tilde{\ell}_4$) be a path in $\dot{\mathfrak{M}}(i)$ from p_2 to q_2 (resp. q_1 to p_1) missing other double points:

5.15 Figure.

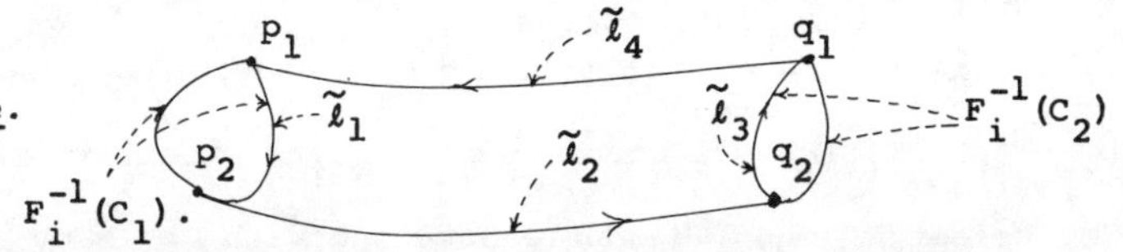

Then $\tilde{\ell}_1\tilde{\ell}_2\tilde{\ell}_3\tilde{\ell}_4$ is null-homotopic relative to $\{p_1\}$ since it misses $b\mathfrak{M}(i)$ and $k \geq 3$; setting $\ell_j = F_i(\tilde{\ell}_j)$, $j = 1,\ldots,4$, then $C_1 = \ell_1$, $C_2 = \ell_3$ and $\ell_1\ell_2\ell_3\ell_4 \simeq 1$ relative to $\{p\}$. Let $\tilde{\ell}_5$ be a path in $\bar{F}_i^{-1}(E)$ from p_1 to q_1 and let $\ell_5 = \bar{F}_i(\tilde{\ell}_5)$. (See (5.16).) Since

5.16 Figure

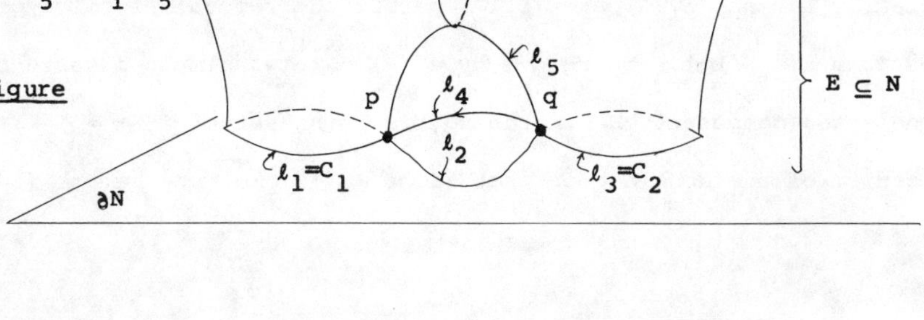

$\mathcal{I}_5 \subseteq \bar{\mathfrak{M}}(i) - b\bar{\mathfrak{M}}(i)$, the self-intersection set of $\bar{F}_i$ is at most 2-dimensional and $k \geq 4$, there is a 2-disc $\tilde{D}_1 \subseteq \bar{\mathfrak{M}}(i)$ such that $\partial\tilde{D}_1 = \tilde{\ell}_2\tilde{\ell}_5^{-1}$ and $\dot{\tilde{D}}_1$ meets the preimages of no double points. Then $D_1 = \bar{F}_i(\tilde{D}_1)$ is an imbedded 2-disc in N with $\partial D_1 = \ell_2\ell_5^{-1}$. Similarly find D_2 imbedded in N with $\partial D_2 = \ell_5\ell_4$. Putting D_1 and D_2 together along ℓ_5 gives $\theta \in \pi_2(N,\partial N)$. Since $\pi_1(\partial N) \overset{\cong}{\to} \pi_1(N)$, $\ell_2\ell_4 \simeq 1$, relative to $\{p\}$. Thus $1 \simeq \ell_1\ell_2\ell_3\ell_4$ $\simeq \ell_1\ell_2\ell_3\ell_2^{-1} \simeq \ell_1(\ell_2\ell_3\ell_2^{-1})$, so $\ell_1 \simeq \ell_1^{-1} \simeq \ell_2\ell_3\ell_2^{-1}$, all relative to $\{p\}$. Since $\ell_1 = C_1$ and $\ell_3 = C_2$, this shows that with respect to any choice of reference paths for F_i and C_1, C_2, C_1 and C_2 contribute the same $g \in \pi_1$ to $\beta(F_i)$.

5.17 <u>Lemma</u>. The boundary of a given connected component E of the self-intersection of $\bar{F}_i\colon \bar{\mathfrak{M}}(i) \to N$ contains an even number of type (b) circles.

<u>Proof</u>. By Lemma (5.14) and Prop. (5.10)(iii), if two type (b) circles occur in E, they can be cancelled. To obtain a contradiction, suppose only one type (b) circle C occurs in ∂E. All other circles in ∂E are then of type (a), which means they are covered by two circles from $\bar{\mathfrak{M}}(i)$. $\bar{F}_i^{-1}(E) \to E$ is a two-sheeted covering of 2-manifolds-with-boundary. Add discs to E and $\bar{F}_i^{-1}(E)$ along all boundary components except those corresponding to C to get $E'' \to E'$, a double covering with $\partial E' = C$, $\partial E'' = C''$ (a circle) where $C'' \to C$ is the non-trivial double covering. But if $E' \to RP^\infty$ classifies $E'' \to E'$, then $\partial E' \to E' \to RP^\infty$ represents the generator of $\pi_1 RP^\infty$ while $\partial E'$ is null-homologous in E', hence in RP^∞. Since $\pi_1 RP^\infty = \mathbb{Z}_2$ is abelian, this is impossible.

Clearly Lemmas (5.14) and (5.17) permit the removal of all type (b) circles from $F(\mathcal{C} - b\mathcal{C})$. To remove type (a) circles simply observe that each connected component E in the double point set of $\bar{F}\colon \bar{\mathcal{C}} \to N$ is a null-bordism of the double point set ∂E of $F\colon \mathcal{C} \to \partial N$. By Remark (5.11) (b), F is regularly homotopic to an imbedding. For details see [HQ, Theorem 3.5].

Before going on to do surgery on $F(\mathcal{C}) = F(\mathcal{C}^k_\mu)$ observe that we have proved the following result.

5.18 <u>Proposition</u>. Let $(g;b): (N^{2k}, \partial N^{2k}; \nu_N) \to (Y,X;\xi)$ be a local surgery problem, where $k \geq 4$ and the non-vanishing kernels appear in the short exact sequence

5.19 $$K_k(N^{2k}, \partial N^{2k}) \rightarrowtail^{\partial} K_{k-1}(\partial N^{2k}) \twoheadrightarrow K_{k-1}(N^{2k}).$$

Then the $(-1)^k$-form $(K_{k-1}(\partial N^{2k}), \varphi, \psi)$ defined by linking and self-linking is a kernel (2.11)(a) with subkernel $\mathrm{im}\{\partial: K_k(N^{2k}, \partial N^{2k}) \rightarrowtail K_{k-1}(\partial N^{2k})\}$.

<u>Proof</u>. The commutative diagram

$$\begin{array}{ccc} H_k(\tilde{\mathcal{C}}^{k+1}_\mu, \tilde{\mathcal{C}}^k_\mu) & \xrightarrow{\cong} & H_{k-1}(\tilde{\mathcal{C}}^k_\mu) \\ \bar{F}_* \downarrow \cong & & \downarrow F_* \\ K_k(N^{2k}, \partial N^{2k}) & \xrightarrow{\partial} & K_{k-1}(\partial N^{2k}) \end{array}$$

shows $\mathrm{im}(\partial) = \mathrm{im}(F_*)$. Since F is an imbedding its image is totally isotropic under φ and ψ by (4B.33).

Returning to the proof of (5.3) let $\Gamma: V \to X \times I$ be the bordism for surgery on $F(\mathcal{C}) \subseteq \partial N$ (4B.53) where $\partial V = \partial N \cup \partial_+ N$. We study the following braid of exact sequences to prove $K_*(\partial_+ N) = 0$:

5.20 <u>Figure</u>

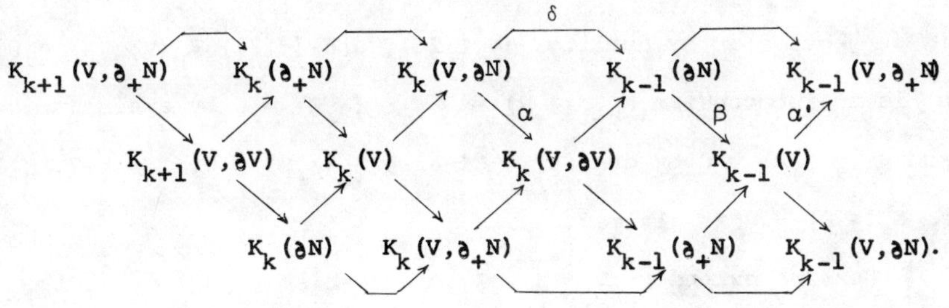

By (4B.46), there is a p.l. homeomorphism

$(V,\partial N) \approx (\partial N \times I \cup R(\bar{\mathcal{C}}'),\partial N \times \{0\})$ where $\bar{\mathcal{C}}' = \bar{\mathcal{C}}'^{k+1}_{\mu}$ is a relative conglomerate constructed according to μ, $R(\bar{\mathcal{C}}') \approx (n\bar{\mathcal{C}}') \times D^{k-2}$ and the union is taken along $R(F(\mathcal{C}))$, the regular neighborhood of $F(\mathcal{C})$ in $\partial N \times \{1\}$ and $R(\mathcal{C}') \subseteq R(\bar{\mathcal{C}}')$, $R(\mathcal{C}') = n\mathcal{C}' \times D^{k-2}$. By excision

5.21
$$K_*(V,\partial N) \xrightarrow{\cong} K_*(R\bar{\mathcal{C}}',R\mathcal{C}')$$

so β in (5.20) is surjective by (4B.25). By duality, $K_{k-1}(N) \cong K_k(N,\partial N)^{\wedge}$ so

5.22
$$|K_{k-1}(N)| = |K_k(N,\partial N)|,$$

where $|T|$ denotes the number of elements in T. The following diagram commutes by construction:

5.23 Figure
$$\begin{array}{ccccc}
K_k(\bar{\mathcal{C}}',\mathcal{C}') & \xrightarrow[\cong]{\partial_1} & K_{k-1}(\mathcal{C}') \equiv K_{k-1}(\mathcal{C}) & \xleftarrow[\cong]{\partial_2} & K_k(\bar{\mathcal{C}},\mathcal{C}) \\
\text{excision}\downarrow\cong & & \downarrow F_* & \cong & \downarrow \bar{F}_* \\
K_k(V,\partial N) & \xrightarrow{\delta} & K_{k-1}(\partial N) & \xleftarrow{\partial} & K_k(N,\partial N)
\end{array}$$

where ∂_1 and ∂_2 are isomorphisms by (4B.25). Thus cok $\delta \cong$ cok ∂ and δ is injective; by (5.20), α is thus injective. By duality α' may be identified with $\alpha^{\wedge}$, so (2.4)(c) implies α' is surjective. If we can show $|K_{k-1}(V)| = |K_{k-1}(V,\partial_+N)|$, then α' is an isomorphism. Indeed, since $K_{k-1}(V,\partial_+N) \cong K_k(V,\partial N)^{\wedge}$, $|K_{k-1}(V,\partial_+N)| = |K_k(V,\partial N)| = |K_k(N,\partial N)|$ (by (5.23)) $= |K_{k-1}(N)|$ (by (5.22)) $= |\text{cok }\partial| = |\text{cok }\delta| = |K_k(V)|$ (since β is surjective). By duality, (4.B.25), and (5.21) $K_k(V,\partial_+N) = 0$; since α' is an isomorphism $K_{k-1}(\partial_+N) = 0$. $K_k(\partial_+N) = 0$ by a similar argument and $K_{k-2}(\partial_+N) = 0$ by duality. Clearly $K_i(\partial_+N) = 0$, $i \neq k-2, k-1, k$, so $K_*(\partial_+N) \equiv 0$ as claimed.

Next we extend $\Gamma: V \to X \times I$ to a bordism $\Psi: (W,N,N_+,V) \to (Y \times I, Y \times \{0\}, Y \times \{1\}, X \times I)$, where $\partial W = N \cup V \cup N_+$,

$V \cap (N \cup N_+) = \partial N \cup \partial_+ N$, and $\partial N_+ = \partial_+ N$. Let $\gamma\colon \partial N \times [a,b] \overset{\approx}{\to} N$ be a collar neighborhood for ∂N in N with $\gamma | \partial N \times \{a\} = \mathrm{id}\colon \partial N \to \partial N$. Using the collar, extend $R(F\mathcal{C}) \subseteq \partial N$ to $R(F\mathcal{C}) \times [a,b] \subseteq N$ and let

5.24 $$W = N \times I \cup R(\bar{\mathcal{C}}') \times [a,b]$$

where the union is taken over $R(F\mathcal{C}) \times [a,b] \subseteq N \times \{1\}$ and $R(\mathcal{C}') \times [a,b] \subseteq R(\bar{\mathcal{C}}') \times [a,b]$. The diagram

$$\begin{array}{ccc} \mathcal{C} = \mathcal{C}^{k}_{\mu} & \xrightarrow{\ F\ } & \partial N \\ \big\downarrow & & \big\downarrow{\scriptstyle g|\partial N} \\ \mathcal{C}^{-k+1}_{\mu} & \longrightarrow & X \end{array}$$

(a boundary of the square in (4B.26)) allows us to extend Γ to Ψ as above.

Now $K_*(W) \otimes \mathbb{Q}\pi \equiv 0$ because of (5.24), (4B.25), and the fact that $K_*(N) \otimes \mathbb{Q}\pi \equiv 0$. Γ is the bordism for surgery on a conglomerate, so $K_*(V) \otimes \mathbb{Q}\pi \equiv 0$. Thus by Definition (1.12), Ψ is a local bordism and by (1.18) g_+ is a local surgery problem (we are omitting bundle data from the notation). Since $g_+ | \partial N_+$ is a homotopy equivalence, the proof of (5.3) is completed by the following two results. (In (5.26), the boundaries, if present, are assumed to be mapped by a homotopy equivalence. Thus they are omitted from the notation.)

5.25 Lemma. $K_k(N_+) \in \mathfrak{D}^1_F$ and is generated by the images of disjoint, framed imbedded k-spheres.

5.26 Proposition. Let $(h;b)\colon (P,\nu_P) \to (Z,\xi)$ be a highly connected local surgery problem. $(K_i(P) = 0,\ i \neq k,\ k-1)$ where $K_k(P) \in \mathfrak{D}^1_F$ and is generated by the images of disjoint, framed imbedded k-spheres. Then $(h;b)$ is locally bordant to a homotopy equivalence.

Proof (of (5.25)). By construction there is a homeomorphism

5.27 $$(N_+,\partial N_+) \approx (N \cup R(\bar{\mathcal{C}}'),(\partial N - \mathring{R}(F\mathcal{C})) \cup \partial' R(\bar{\mathcal{C}}'))$$

where $\partial' R(\bar{\mathcal{C}}') = \partial R(\bar{\mathcal{C}}') - \mathring{R}(\mathcal{C}')$ and the first union is taken by identifying $R(F\mathcal{C})$ with $R(\mathcal{C}')$. Identifying left and right sides in (5.27) yields excisions

5.28(a) $$(N,\partial N) \hookrightarrow (N_+,\partial N_+ \cup R(\bar{\mathcal{C}}'))$$

(b) $$(R(\bar{\mathcal{C}}'),R(\bar{\mathcal{C}}') \cap \partial N_+) \hookrightarrow (\partial N_+ \cup R(\bar{\mathcal{C}}'),\partial N_+) .$$

Consider the following (horizontal) portion of the exact sequence of the triple $(N_+,\partial N_+ \cup R(\bar{\mathcal{C}}'),\partial N_+)$:

$$\begin{array}{ccccccc}
K_k(\partial N_+ \cup R(\bar{\mathcal{C}}'),\partial N_+) & \xrightarrow{i} & K_k(N_+,\partial N_+) & \xrightarrow{j} & K_k(N_+,\partial N_+ \cup R(\bar{\mathcal{C}}')) & \xrightarrow{\partial} & K_{k-1}(\partial N_+ \cup R(\bar{\mathcal{C}}'),\partial N_+) \\
\uparrow \cong & & & & \uparrow \cong & & \uparrow \cong \\
K_k(R(\bar{\mathcal{C}}'),R(\bar{\mathcal{C}}') \cap \partial N_+) & & & & K_k(N,\partial N) & & K_{k-1}(R(\bar{\mathcal{C}}'),R(\bar{\mathcal{C}}') \cap \partial N_+)
\end{array}$$

5.29 <u>Figure</u>

By construction $K_k(N,\partial N)$, and hence also $K_k(N_+,\partial N_+ \cup R(\bar{\mathcal{C}}'))$, is generated by the relative classes $[\bar{F}(b_i\bar{\mathcal{C}})]$, $i = 1,\ldots,n$. (See (5.13) and (5.28(a).) Since $\partial[\bar{F}(b_i\mathcal{C})] = [F(b_i\mathcal{C})]$ and the boundary of $b_i\bar{\mathcal{C}}'$ is $F(b_i\mathcal{C})$ in $\partial N_+ \cup R(\bar{\mathcal{C}}')$, $\partial = 0$ in (5.29). Since $K_k(\partial N_+ \cup R(\bar{\mathcal{C}}'),\partial N_+) \cong K_k(R(\bar{\mathcal{C}}'),R(\bar{\mathcal{C}}') \cap \partial N_+) \cong K_{k-1}(R(\bar{\mathcal{C}}'),R(\bar{\mathcal{C}}') \cap \partial N)^\wedge$ (duality) $\cong K_{k-1}(\bar{\mathcal{C}}',\mathcal{C}')$ (since $R(\bar{\mathcal{C}}') \cap \partial N = R(\mathcal{C}')$) and the last term is zero by (4B.25), we find that j is an isomorphism in (5.29). Thus the homology classes of the framed imbedded k-spheres $\bar{F}(b_i\bar{\mathcal{C}}) \cup (-b_i\bar{\mathcal{C}}')$ generate $K_k(N_+,\partial N_+) = K_k(N_+)$, $i = 1,\ldots,n$, since the classes $[\bar{F}_i(b_i\bar{\mathcal{C}})]$ generate $K_k(N_+,\partial N_+ \cup R(\bar{\mathcal{C}}'))$.

<u>Proof</u> (of (5.26)). Denote by $H:W \to Z \times I$ the bordism of h corresponding to surgery on a set of generators of $K_{k-1}(P)$, and let $\partial W = P \cup P'$. The non-zero kernels are contained in the following braid diagram

5.30 <u>Figure</u>

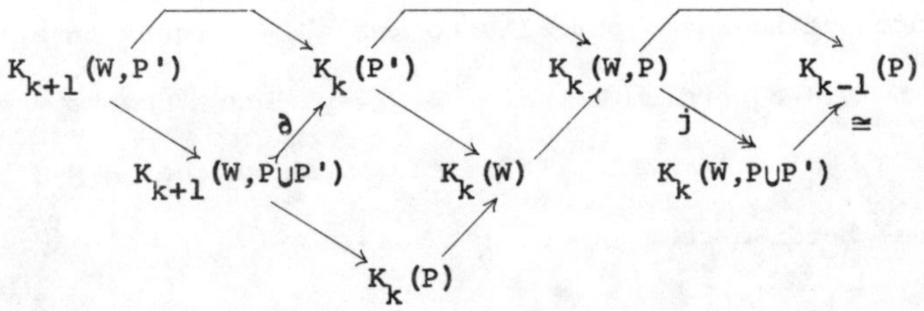

Since $K_k(P')$ is $\mathbb{Z}\pi$-free, $K_{k+1}(W,P\cup P')$ is $\mathbb{Z}$-torsion free. Since $K_k(P) \in \mathfrak{D}_F^1$ and $K_{k+1}(W,P)$ is $\mathbb{Z}\pi$-free, it follows from the short exact sequence

5.31 $$K_{k+1}(W,P') \rightarrowtail K_{k+1}(W,P\cup P') \twoheadrightarrow K_k(P)$$

and Schanuel's Lemma [Ma,p.101] that $K_k(W,P\cup P')$ is stably free. By adding the classes of sufficiently many null-homotopic spheres to the list of generators of $K_{k-1}(P)$ on which the surgery was done, we may suppose $K_{k+1}(W,P\cup P')$ free over $\mathbb{Z}\pi$.

Let $S_1,\ldots,S_m \subseteq P$ denote disjoint, framed imbedded k-spheres, disjoint from the generators of $K_{k-1}(P)$, such that $\{[S_1],\ldots,[S_m]\}$ generates $K_k(P)$. From (5.31) it follows that $\{c_1,\ldots,c_n,[S_1\times I],\ldots,[S_m\times I]\}$ is a generating set for $K_{k+1}(W,P\cup P')$ where the c_i are homology classes of the "transverse discs" in the surgery carried out above. Hence the image of $\partial\colon K_{k+1}(W,P\cup P') \to K_k(P')$ is a free $\mathbb{Z}\pi$-module annihilated by the intersection and self-intersection forms on $K_k(P')$. We claim $\text{im}(\partial)$ is a subkernel. Indeed from (5.30) we extract the short exact sequence

$$\text{im}(\partial) \rightarrowtail K_k(P') \twoheadrightarrow \ker(j).$$

Since $K_k(P) \in \mathfrak{D}_F^1$ and (by duality) $K_{k-1}(P) \cong K_k(P)^\wedge$, it follows from (2.15) that $K_{k-1}(P) \in \mathfrak{D}_F^1$. From $K_k(W,P\cup P') \cong K_{k-1}(P)$ and the argument above with Schanuel's Lemma, we can assume $\ker(j)$ is $\mathbb{Z}\pi$-free. It is easy to see that the $\mathbb{Z}\pi$-ranks of $\text{im}(\partial)$ and $\ker(j)$ are equal, so $\text{im}(\partial)$ is a subkernel. By [W2,5.3] (or rather its proof), there is another subkernel L of

$K_k(P')$ projecting isomorphically to $\ker(j)$. Surgery on a free generating set of L via a bordism $H'\colon W' \to Z \times I$ yields a homotopy equivalence $h''\colon P'' \to Z$, $\partial W' = P' \cup P''$. It is easily shown that $H \cup H'\colon W \cup W' \to Z \times I$ is a local bordism from h to h''.

Chapter Six: Local Surgery in the Odd-dimensional Case.

The goal of this chapter is to identify the bordism group of local surgery problems ${}^{1}L^{\mathbb{Q}/\mathbb{Z}}_{2k+1}(K)$ (1.5) with the "reduced" Grothendieck group of ϵ-forms $L^{\epsilon}_{0}(\mathbb{Q}\pi/\mathbb{Z}\pi)$ (2.13) where $\epsilon = (-1)^{k+1}$ and $\pi = \pi_1 K$ is finite. The precise statement is:

6.1 Theorem. Let K be a finite connected complex with $\pi_1 K$ finite. Then there is a natural isomorphism of groups for any $k \geq 4$,

$$ {}^{1}L^{\mathbb{Q}/\mathbb{Z}}_{2k+1}(K) \xrightarrow{\cong} L^{(-1)^{k+1}}_{0}(\mathbb{Q}[\pi_1 K]/\mathbb{Z}[\pi_1 K]). $$

The meaning of "naturality" in (6.1) is evident. The main point is to show that the obstruction to odd-dimensional local surgery lies in $L^{\epsilon}_{0}(\mathbb{Q}\pi/\mathbb{Z}\pi)$ and that all elements of the latter group so occur. This is Prop. (6.2). Given this, the (π-π)-Theorem (5.3), and the results of Chapter One, the derivation of (6.1) is formal and left to the reader. (For details see [W2,9.4.1].) Functoriality of $L^{\epsilon}_{0}(\mathbb{Q}\pi/\mathbb{Z}\pi)$ with respect to group homomorphisms $\pi \to \pi'$ is treated in [Ha]; from this follows naturality in (6.1).

Let $(g;b)\colon (N^{2k+1}, \partial N^{2k+1}; \nu_N) \to (Y,X;\xi)$ be a highly-connected local surgery problem ($K_i(N) = 0$, $i \neq k$) where $k \geq 4$, $g_{\#}\colon \pi_1 N \xrightarrow{\cong} \pi_1 Y$ (finite groups), and $g|\partial N$ is a homotopy equivalence. As usual, $g|\partial N$ will be fixed in all arguments and therefore is often ignored in the notation. By (2.10) and (4A.37) there is an ϵ-form $(K_k(N),\varphi,\psi)$ where $\epsilon = (-1)^{k+1}$ and φ (resp. ψ) is the linking form (resp. self-linking form) on $K_k(N)$. Let $[K_k(N),\varphi,\psi]$ denote the class of $(K_k(N),\varphi,\psi)$ in $L^{\epsilon}_{0}(\mathbb{Q}\pi/\mathbb{Z}\pi)$, where $\pi = \pi_1 N$.

6.2 Proposition. Let $k \geq 4$, and let π be a finite group. A highly-connected local surgery problem $(g;b)\colon (N^{2k+1},\partial N^{2k+1};\nu_N) \to (Y,X;\xi)$ is locally bordant to a homotopy equivalence (relative to $g|\partial N$) if and only if $[K_k(N),\varphi,\psi] = 0$ in $L_0^{(-1)^{k+1}}(\mathbb{Q}\pi/\mathbb{Z}\pi)$, where $\pi = \pi_1 Y = \pi_1 N$. Given a $(-1)^{k+1}$-form (S,φ,ψ), there is a highly connected local surgery problem $(h;b)\colon (P^{2k+1};\nu_P) \to (Z;\eta)$ such that $[K_k(P),\zeta,\xi] = [S,\varphi,\psi]$, where ζ and ξ are the linking and self-linking forms on $K_k(P)$.

Proof. We omit bundle and dimension notation. Let g be locally bordant to a homotopy equivalence $g'\colon N' \to Y$ via the bordism $G\colon W \to Y \times I$ where $\partial W = N \cup N'$ and $g = G|N$, $g' = G|N'$. By (1.20) (or rather its proof) we may do local surgery on the interior of W until the non-zero kernels of $(G, g \cup g')$ occur in the exact sequence

$$K_{k+1}(W) \rightarrowtail K_{k+1}(W,N \cup N') \longrightarrow K_k(N \cup N') \longrightarrow K_k(W) \twoheadrightarrow K_k(W,N \cup N').$$

By (5.1) this can be reduced to

6.3 $$K_{k+1}(W,\bar{N} \cup N') \rightarrowtail K_k(\bar{N} \cup N') \twoheadrightarrow K_k(W)$$

where $\bar{N}$ is the result of surgery on a conglomerate $\bar{F}(\mathcal{C}^k_\mu) \subseteq N$ and $\bar{F}\colon \mathcal{C}^{k+1}_\mu \to N$ is a framed imbedding. By (4B.60), $(K_k(\bar{N}),\bar{\varphi},\bar{\psi})$ is isometric to $(K_k(N),\varphi,\psi) \perp \mathcal{H}(S)$ where $S = \operatorname{cok}(\mu)$ and $\mathcal{H}(S)$ is the hyperbolic form on S ((2.11)(b)). By (5.18), $(K_k(\bar{N} \cup N'),(\bar{\varphi} \cup \varphi'),(\bar{\psi} \cup \psi'))$ (with obvious notation) is a kernel; it is isometric to $(K_k(\bar{N}),\bar{\varphi},\bar{\psi})$ since $K_*(N') \equiv 0$. Hence $[K_k(N),\varphi,\psi] = 0$.

Conversely, suppose given $(g;b)$ as in the statement of (6.2) with $[K_k(N),\varphi,\psi] = 0$. By (2.22) there is a hyperbolic ϵ-form $\mathcal{H}(U)$ and an isometry $(K_k(N),\varphi,\psi) \perp \mathcal{H}(U) \cong (T,\varphi',\psi')$, where (T',φ',ψ') is a kernel. By (4B.58) and (4B.60) we may assume $(K_k(N),\varphi,\psi)$ is itself a kernel. Thus to show that $(g;b)$ is locally bordant to a homotopy equivalence, it remains to prove

6.4 Proposition. Suppose given a highly connected local surgery problem $(g;b)$: $(P^{2k+1};\nu_P) \to (Z;\eta)$ where $(K_k(P^{2k+1}),\varphi,\psi)$ is a kernel. Then $(g;b)$ is locally bordant to a homotopy equivalence.

Proof. Let $E \rightarrowtail F \twoheadrightarrow J$ be a short free resolution of a subkernel J of $K_k(P)$ and let $\{x_1,\ldots,x_n\} \subseteq K_k(P)$ be the image of a $\mathbb{Z}\pi$-basis of F. Do surgery on the x_i via a bordism $G: V \to Z \times I$ with $\partial V = P' \cup P$ and $g' = G|P'$. Consider the braid diagram

6.5 Diagram

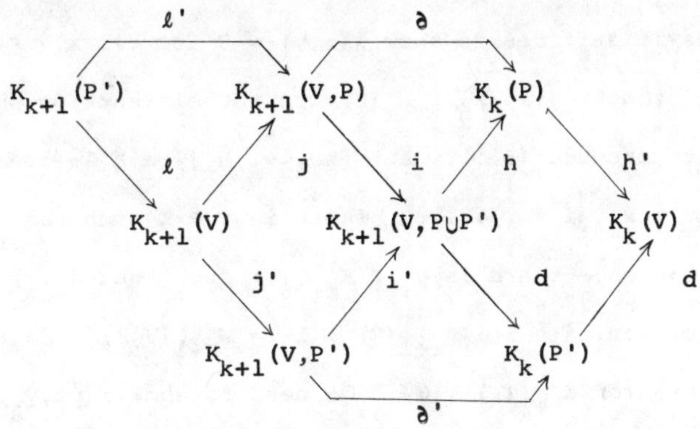

The proof of (6.4) is decomposed into three lemmas.

6.6 Lemma. In Diagram (6.5), $j' = 0$, ℓ is an isomorphism, ℓ' and ∂' are injective, and h,h',d, and d' are surjective.

6.7 Lemma. In Diagram (6.5), $K_k(P')$ is $\mathbb{Z}\pi$-free. Surgery on a basis via the bordism G': $V' \to Z \times I$, where $\partial V' = P'' \cup P'$, yields a homotopy equivalence $G|P'' = g''$: $P'' \to Z$.

6.8 Lemma. Let $W = V \cup V'$ and $H = G \cup G'$. Then H: $(W,P,P'') \to (Z \times I, Z \times \{0\}, Z \times \{1\})$ is bordant to a local surgery problem relative to $H|P \cup P'' = g \cup g''$.

Proof (of (6.6)). To prove $j' = 0$, it suffices to show $i'j' = 0$ since i' is injective ($K_{k+1}(P) = 0$). Recall from (2.2) that τM denotes the torsion part and $fM = M/{}^{T}M$ the free "part" of a $\mathbb{Z}\pi$-module M. Since $K_{k+1}(V,P')$ is free, $\mathrm{im}\,(i'j') \cap {}^{T}K_{k+1}(V,P \cup P') = (0)$, so it suffices to prove that the composition

6.9 $$K_{k+1}(V) \longrightarrow K_{k+1}(V,P') \longrightarrow K_{k+1}(V,P \cup P') \longrightarrow fK_{k+1}(V,P \cup P')$$

is the zero map. Duality induces an isomorphism $fK_{k+1}(V,P \cup P') \overset{\cong}{\rightarrow} \overline{K_{k+1}(V)}$ and its composition with (6.9) yields a map $K_{k+1}(V) \rightarrow \overline{K_{k+1}(V)}$ which is well-known to be the adjoint of the intersection form $\lambda\colon K_{k+1}(V) \times K_{k+1}(V) \rightarrow \mathbb{Z}\pi$. Thus it suffices to show $\lambda(x,y) = 0$ for all $x,y \in K_{k+1}(V)$.

Recall that $x_1,\ldots,x_n \in K_k(P)$ are the elements on which surgery has been done to produce (6.5). Let $\{h_1,\ldots,h_n\}$ be a $\mathbb{Z}\pi$-basis for $K_{k+1}(V,P)$, so that $\partial h_i = x_i$, $i = 1,\ldots,n$. There is $r \in \mathbb{Z}$ such that $rx_i = 0$, $i = 1,\ldots,n$; hence there is $y_i \in K_{k+1}(V)$ such that $j(y_i) = rh_i$. Since $\mathrm{im}(\partial)$ is torsion, $j \otimes \mathbb{Q}\pi\colon K_{k+1}(V) \otimes \mathbb{Q}\pi \overset{\cong}{\rightarrow} K_{k+1}(V,P) \otimes \mathbb{Q}\pi$, so $\{y_1,\ldots,y_n\}$ is a $\mathbb{Q}\pi$-basis for $K_{k+1}(V) \otimes \mathbb{Q}\pi$. We need to show $\lambda(y_i,y_j) = 0$, $i,j = 1,\ldots,n$.

Since $rx_i = 0$, (3.17) and (4A.21) furnish framed immersions $F_i\colon \mathfrak{M}_r^{k+1} \rightarrow P$ with $[F(b\mathfrak{M}_r^{k+1})] = x_i$ where $[X]$ denotes the homology class of the cycle X. Since for all i and j, $\varphi(x_i,x_j) = 0 = \psi(x_i)$ we may assume by (4A.39) that

6.10 $$F_i'(b\mathfrak{M}_r^{k+1}) \cap F_j(\mathfrak{M}_r^{k+1}) = \emptyset,$$

where F_i' denotes F_i pushed a non-zero distance along the first vector of its complementary framing. Let $h_i = [H_i]$ where H_i is the core $(k+1)$-disc of handle killing x_i. Clearly, $y_i = [F_i(\mathfrak{M}_r^{k+1}) \cup (-rH_i)]$. To put these cycles in $\mathring{V}$, push the images of the F_i into $\mathring{V}$ (each a different distance) along the collar direction of P in V, obtaining $F_i''\colon \mathfrak{M}_r^{k+1} \rightarrow V$. Since the H_i are disjoint, the only possible intersections among the cycles cycles $F_i''(\mathfrak{M}_r^{k+1}) \cup (-rH_i)$ must arise from intersections of the form

$F''_i(b\mathfrak{m}_r^{k+1}) \cap F''_j(\mathfrak{m}_r^{k+1})$, since $\partial H_i = F_i(b\mathfrak{m}_r^{k+1})$. By (6.10) these intersections are empty, so we have shown $j' = 0$. The rest of (6.6) is left to the reader.

<u>Proof</u> (of (6.7)). Since ℓ is an isomorphism (Diagram (6.5) and Lemma (6.6)), h induces an isomorphism $\text{im}(i) \xrightarrow{\cong} \text{im}(\partial)$, which by construction is the subkernel J. Thus $\text{im}(i) \subseteq \tau K_{k+1}(V,P \cup P')$ and by duality $\tau K_{k+1}(V,P \cup P') \cong K_k(V)^{\wedge}$ ($K_k(V)$ is clearly torsion). Since $|J|^2 = |K_k(P)|$ ((2.12)(b)) the short exact sequence

$$J = \text{im}(\partial) \rightarrowtail K_k(P) \twoheadrightarrow K_k(V)$$

shows $|J| = |K_k(V)|$. Using the fact that $|K_k(V)| = |K_k(V)^{\wedge}|$ ((2.12)(b)), the above results give

$$|\text{im}(i)| = |\text{im}(\partial)| = |J| = |K_k(V)| = |\tau K_{k+1}(V,P \cup P')|.$$

Hence $\text{im}(i) = \tau K_{k+1}(V,P \cup P')$.

Next consider the commutative diagrams

6.11 Figure

$$\begin{array}{ccc} \text{im}(i) = \tau K_{k+1}(V,P\cup P') & \xrightarrow{\tau(d)} & \tau K_k(P') \\ \downarrow & & \downarrow \\ K_{k+1}(V,P\cup P') & \xrightarrow{d} & K_k(P') \\ \downarrow & & \downarrow \\ fK_{k+1}(V,P\cup P') & \xrightarrow{f(d)} & fK_k(P') \end{array}$$

(a)

$$\begin{array}{ccc} fK_{k+1}(V,P\cup P') & \xrightarrow{f(d)} & fK_k(P') \\ \cong \downarrow & & \cong \downarrow \\ \overline{K_{k+1}(V)} & \xrightarrow{\bar{\ell}} & \overline{K_k(P')} \end{array}$$

(b)

where the vertical sequence of (6.11)(a) are short exact, $\tau(d)$ and $f(d)$ are the induced maps, and the verticals of (6.11)(b) are duality isomorphisms ($K_{k+1}(V)$ and $K_{k+1}(P')$ are torsion-free). We observed above that ℓ is an isomorphism, so $f(d)$ is an isomorphism ((6.11)(b)). By (6.11)(a), $\tau(d)$ is a surjection while it annihilates $\text{im}(i)$ ($d \circ i = 0$). Thus $\tau K_k(P') = 0$.

The exact sequence

$$K_{k+1}(V) \rightarrowtail K_{k+1}(V,P) \xrightarrow{\partial} \twoheadrightarrow \mathrm{Im}(\partial) = J,$$

the assumption that $J \in \mathfrak{D}^1_F$, and Schanuel's Lemma [Ma, p. 101], show $K_{k+1}(V)$ is stably free. By the usual argument we may assume $K_{k+1}(V)$ is free (add zero classes to the list $\{x_1,\ldots,x_n\}$ of classes killed by surgery). By (6.11)(b) and the fact that $\tau K_k(P') = 0$, $K_k(P')$ is $\mathbb{Z}\pi$-free. The second statement in (6.7) is left to the reader.

<u>Proof</u> of (Lemma (6.8)). The non-zero kernels for $G'\colon V'' \to Z \times I$ appear in Figure (6.12) and the maps are as labeled:

6.12 Figure

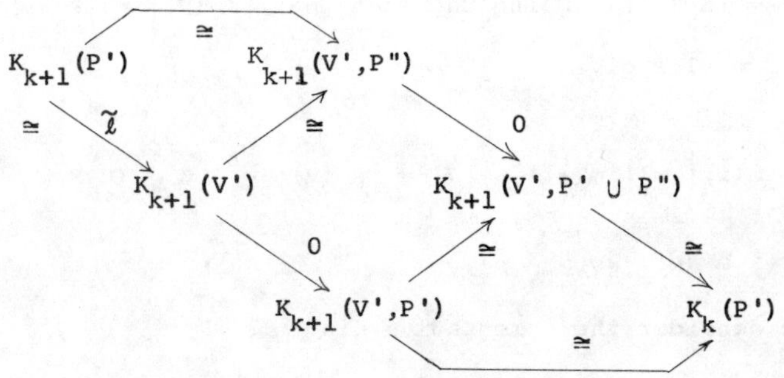

The relevant part of the Mayer-Vietoris sequence for $(V \cup V'; V, V'; V \cap V')$ is

$$\text{6.13}\quad \begin{array}{ccccccccccc} K_{k+2}(V\cup V') & \to & K_{k+1}(V\cap V') & \to & K_{k+1}(V)+K_{k+1}(V') & \to & K_{k+1}(V\cup V') & \to & K_k(V\cap V') & \to & K_k(V) \\ & & \| & \nearrow_{\ell+\tilde{\ell}} & & & & & \| & \nearrow_{d'} & \\ & & K_k(P') & & & & & & K_k(P') & & \end{array}$$

Since ℓ (from (6.5)) and $\tilde{\ell}$ are isomorphisms, and d' is surjective, $K_i(V \cup V') = 0$, $i = k+2,k$. Hence $K_i(V \cup V') = 0$, $i \neq k+1$. Since $K_k(V) \otimes \mathbb{Q}\pi = 0$ and tensoring (6.13) with $\mathbb{Q}\pi$ preserves exactness (1.16), a dimension count shows $K_{k+1}(V \cup V') \otimes \mathbb{Q}\pi \cong (\mathbb{Q}\pi)^{2n}$; $K_{k+1}(V) \to K_{k+1}(V \cup V')$ is an injection of a free $\mathbb{Z}\pi$-module of rank n, whose intersection form is trivial (see proof of (6.6)). Surgery on a basis for the image of

$K_{k+1}(V)$ completes the proof of (6.8).

6.14 Remark. Given our contention that local surgery is like ordinary surgery, (6.4) ought to have a straightforward proof using conglomerates (compare [W2,p.51]). Indeed, the absolute version of (4B.26) furnishes a framed immersion $F: \mathcal{C}_\mu^{k+1} \to P^{2k+1}$ such that $\operatorname{im}\{F_*: H_k(\tilde{\mathcal{C}}_\mu^{k+1}) \to K_k(P)\}$ $= J$ (the subkernel). By (4B.36) we may assume $F(\mathring{\mathcal{C}}_\mu^{k+1}) \cap F(b\mathcal{C}_\mu^{k+1}) = \emptyset$. If we could further modify F to an imbedding, then it is easy to show that surgery on $F(\mathcal{C}_\mu^{k+1})$ would produce a homotopy equivalence. It is probably possible to so modify F, but this means removing its remaining (circle) self-intersections, so a direct argument seems difficult.

To complete the proof of (6.2) it remains to construct a local surgery problem with given surgery obstruction. This result is due to Connolly [Co.4.4]; our proof is closely related to Connolly's except that our use of conglomerates makes clear the analogy with "plumbing" and is needed in Chapter Seven.

Let M^{2k} be a closed p.l. manifold with $\pi_1 M = \pi$ finite. Let (S,φ,ψ) be a $(-1)^{k+1}$-form with covering (R,τ) (see (2.19)) where

$$(R) = (E \overset{\mu}{\rightarrowtail} F \twoheadrightarrow S) \quad \text{and} \quad \tau: F \times F \longrightarrow \mathbb{Q}\pi$$

are a short free resolution of S and a $(-1)^{k+1}$-hermitian form, respectively. As usual we also think of τ and μ as $(n\times n)$-matrices. By (2.19)(c) the matrix $\tau\mu$ has entries in $\mathbb{Z}\pi$. Let $\bar{F}: \mathcal{C}_\mu^{k+1} \to M^{2k}$ be a framed imbedding (4B.58) and let

$$H: \mathcal{C}_\mu^{k} \times I \longrightarrow M^{2k} \times I$$

be a regular homotopy of $F = \bar{F}|\mathcal{C}_\mu^{k}$ intersecting itself in transverse position such that

6.15 $$H(b_i\mathcal{C}_\mu^{k} \times I) \pitchfork H(\mathfrak{m}(j) \times I) = (\tau\mu)_{ij},$$

where $\mathcal{C}^k_\mu = \bigcup_{i=1}^{n} \mathfrak{m}(i)$ (see (4B.14)). H is constructed by pushing $\mathring{\mathfrak{m}}(j)$ across $b_i\mathcal{C}^k_\mu$ with intersection number $(\tau\mu)_{ij}$, for each i and j; the bockstein of $\mathcal{C}^k_\mu$ is kept fixed. Denoting $F_1 = H|\mathcal{C}^k_\mu \times \{1\}$, $F_1|\mathring{\mathcal{C}}^k_\mu$ intersects itself transversely in isolated points. We will show next (without assuming $F = H|\mathcal{C}^k_\mu \times \{0\}$ extends to $\bar{F}:\bar{\mathcal{C}}^{k+1}_\mu \to M^{2k}$) that F_1 is regularly homotopic to an imbedding (fixing the bockstein).

6.16 <u>Lemma</u>. Given any framed imbedding $F: \mathcal{C}^k_\mu \to M^{2k}$, we may construct a regular homotopy of F, $H: \mathcal{C}^k_\mu \times I \to M^{2k} \times I$ satisfying (6.15) and

$$F_1(\mathring{\mathfrak{m}}(i)) \pitchfork F_1(\mathring{\mathfrak{m}}(j)) = 0, \quad i,j = 1,\ldots,n.$$

Hence, we may assume $F_1 = H|\mathcal{C}^k_\mu \times \{1\}$ is a framed imbedding.

<u>Proof</u>. Fix i and j. For any $\ell = 1,\ldots,n$, pushing $\mathring{\mathfrak{m}}(j)$ across $b_\ell\mathcal{C}^k_\mu$ with intersection number $(\tau\mu)_{\ell j}$ contributes to $F_1(\mathring{\mathfrak{m}}(i)) \pitchfork F_1(\mathring{\mathfrak{m}}(j))$

$$\{H(b_\ell\mathcal{C}^k_\mu \times I)\}\mu_{\ell i} \pitchfork H(\mathfrak{m}(j) \times I) = \bar{\mu}_{\ell i}(\tau\mu)_{\ell j}$$

since $\mathfrak{m}(i)$ abuts to $b_\ell\mathcal{C}^k_\mu$ with multiplicity $\mu_{\ell i}$. Summing over ℓ, the total contribution to $F_1(\mathring{\mathfrak{m}}(i)) \pitchfork F_1(\mathring{\mathfrak{m}}(j))$ is

6.17 $$\overline{\sum_\ell \mu_{\ell i}(\tau\mu)_{\ell j}} = (i,j)\text{-entry in } \bar{\mu}\tau\mu.$$

Likewise, for $m = 1,\ldots,n$, pushing $\mathring{\mathfrak{m}}(i)$ across $b_m\mathcal{C}^k_\mu$ contributes $\sum_m \bar{\mu}_{mj}(\tau\mu)_{mi}$ to $F_1(\mathring{\mathfrak{m}}(j)) \pitchfork F_1(\mathring{\mathfrak{m}}(i))$ or

6.18 $$\overline{(-1)^k \sum_m \bar{\mu}_{mj}(\tau\mu)_{mi}} = (-1)^k \sum_m \overline{(\tau\mu)}_{mi}\mu_{mj}$$

$$= (-1)^k\{(i,j)\text{-entry of } \bar{\mu}\bar{\tau}\mu\}$$

to $F_1(\mathring{\mathfrak{m}}(i)) \pitchfork F_1(\mathring{\mathfrak{m}}(j))$. Since $\tau = (-1)^{k+1}\bar{\tau}$, the sum of (6.17) and (6.18) is zero, whence the lemma.

Applying the Whitney device repeatedly to $F_1|\mathring{\mathcal{C}}^k_\mu$ we may assume F_1 is a framed imbedding. (In the process, circle intersections are created in $H: \mathcal{C}_\mu^k \times I \to M^{2k} \times I$, but these are not important.) Let $h: P^{2k+1} \to M^{2k} \times I$ denote the bordism for surgery on $F_1(\mathcal{C}_\mu^k)$ (in the domain of the (identity) local surgery problem $M \xrightarrow{id} M$) and let $\partial P^{2k+1} = M^{2k+1} \cup M'^{2k+1}$. There is a homeomorphism (4B.46)(ii)

$$(P^{2k+1}, M^{2k}) \approx M^{2k} \times I \cup R(\bar{\mathcal{C}}'^{k+1}_\mu)$$

where $\bar{\mathcal{C}}'^{k+1}_\mu$ is the relative conglomerate constructed according to the matrix μ, $R(\bar{\mathcal{C}}'^{k+1}_\mu)$ is the total space of its trivial normal bundle and $R(\mathcal{C}'^k_\mu) \equiv R(F_1\mathcal{C}^k_\mu)$, the regular neighborhood of $F_1\mathcal{C}^k_\mu$ in M^{2k}. Let us drop superscripts denoting dimension of a manifold.

Using $K_*(M) \equiv 0$ and (4B.46)(iii)

6.19 $$K_i(P) = K_i(P,M) = \begin{cases} S, & i = k \\ 0, & i \neq k \end{cases}.$$

Let $\mathcal{C}''^{k+1}_\mu = \bar{\mathcal{C}}'^{k+1}_\mu \cup H(\mathcal{C}^k_\mu \times I) \cup \bar{F}(\bar{\mathcal{C}}^{k+1}_\mu)$. Then $\mathcal{C}''^{k+1}_\mu$ is a conglomerate constructed according to μ and is framed immersed in P. Consider the diagram

6.20 Diagram
$$\begin{array}{ccc}
H_k(\tilde{\mathcal{C}}''^{k+1}_\mu) & \longrightarrow & K_k(P) \\
\downarrow \cong & & \downarrow \cong \\
H_k(\tilde{\mathcal{C}}''^{k+1}_\mu, H(\tilde{\mathcal{C}}^k_\mu \times I) \cup \bar{F}(\tilde{\mathcal{C}}^{k+1}_\mu)) & \longrightarrow & K_k(P, M \times I) \\
\nwarrow \cong & & \nearrow \cong \ \text{(4B.46)(iii)} \\
 & H_k(\tilde{\bar{\mathcal{C}}}'^{k+1}_\mu, \tilde{\mathcal{C}}'^k_\mu) &
\end{array}$$

where the upper left vertical uses (4B.25)(i) and the fact that $H(\mathcal{C}^k_\mu \times I) \cup \bar{F}(\bar{\mathcal{C}}^{k+1}_\mu)$ is a relative conglomerate. Thus (6.20) yields

6.21 $$H_k(\tilde{\mathcal{C}}''^{k+1}_\mu) \xrightarrow{\cong} K_k(P).$$

By (6.19), (6.20), and (4B.33) the linking and self-linking forms on $K_k(P)$

are

$$\varphi : K_k(P) \times K_k(P) \longrightarrow \mathbb{Q}\pi/\mathbb{Z}\pi$$

$$\psi : K_k(P) \longrightarrow \mathbb{Q}\pi/J_{(-1)^{k+1}}.$$

If $\varphi' : K_k(P,\partial P) \times K_k(P) \to \mathbb{Q}\pi/\mathbb{Z}\pi$ is the nonsingular linking pairing, the diagram

$$\begin{array}{ccc} K_k(P) & \xrightarrow{\;j\;} & K_k(P,\partial P) \\ & \searrow^{\mathrm{Ad}(\varphi)} \quad {}^{\mathrm{Ad}(\varphi')}\swarrow & \\ & K_k(P)^{\wedge} & \end{array}$$

commutes. Since φ and φ' are nonsingular, j is an isomorphism. By construction of P and (6.19) the non-zero kernels for $h: (P,\partial P) \to (M \times I, M \times \{0\} \cup M \times \{1\})$ appear in the exact sequence

$$K_k(\partial P) \rightarrowtail K_k(P) \xrightarrow{\;j\;} K_k(P,\partial P) \twoheadrightarrow K_{k-1}(\partial P).$$

Since j is an isomorphism, $K_*(\partial P) \equiv 0$ and the proof of (6.2) is complete.

Chapter Seven: Local Surgery in the Even-dimensional Case

In this chapter the following theorem is proved. (Recall from (1.5) the definition of ${}^{1}L_n^{\mathbb{Q}/\mathbb{Z}}(K)$)

7.1 Theorem. Let K be a finite connected complex with $\pi_1 K$ finite. Then there is a natural isomorphism of commutative semi-groups for any $k \geq 4$,

$$ {}^{1}L_{2k}^{\mathbb{Q}/\mathbb{Z}}(K) \xrightarrow{\cong} L_1^{\epsilon}(\mathbb{Q}[\pi_1 K]/\mathbb{Z}[\pi_1 K]), \qquad \epsilon = (-1)^k. $$

Here naturality has the obvious meaning and can be verified (once we construct the isomorphism in (7.1)) just as for (6.1). Recall that ${}^{1}L_{2k}^{\mathbb{Q}/\mathbb{Z}}(K)$ has a natural group structure.

7.2 Corollary. $L_1^{\epsilon}(\mathbb{Q}\pi/\mathbb{Z}\pi)$ is an abelian group.

Theorem (7.1) follows formally from the $(\pi-\pi)$-Theorem (5.3), the results in Chapter One, and the following proposition (just as (6.1) follows from (6.2)--see [W2, Chapter 9]).

7.3 Proposition. Let $k \geq 4$ and let π be a finite group. To each highly-connected local surgery problem $(g;b)\colon (N^{2k}, \partial N^{2k}; \nu_N) \to (Y,X;\xi)$ where $g_\#\colon \pi_1 N \xrightarrow{\cong} \pi_1 Y = \pi$, we may associate $\theta(g,b) \in L_1^{(-1)^k}(\mathbb{Q}\pi/\mathbb{Z}\pi)$ so that $\theta(g,b) = 0$ if and only if (g,b) is locally bordant (rel ∂N) to a homotopy equivalence. Given $\theta \in L_1^{(-1)^k}(\mathbb{Q}\pi/\mathbb{Z}\pi)$ there is a highly connected local surgery problem $(h;b)\colon (P^{2k}, \partial P; \nu_P) \to (Z, \partial Z; \xi)$ such that $\theta(h,b) = \theta$.

From now on boundary components (∂N^{2k} and X above) are understood to be mapped by a fixed homotopy equivalence ($g|\partial N^{2k}$) and so are ignored in the notation. Also we assume for the rest of the chapter that k is an integer and $k \geq 4$; N always means N^{2k}. The proof of (7.3) occupies the rest of this chapter; specifically, the proof of the first assertion is in (7.53) and (7.68) and the second is in (7.69). To begin with we construct $\theta(g,b)$.

Suppose given a highly connected local surgery problem $(g,b)\colon (N^{2k},\nu_N) \to (Y,\xi)$ $(K_i(N) = 0,\ i \neq k,\ k-1,\ k \geq 4)$, $\mu \in M_n(\mathbb{Z}\pi,\mathbb{Q}\pi)^{\times}$, and a commutative diagram

7.4
$$\underline{\underline{F}} = \begin{array}{ccc} \mathcal{C}_\mu^k & \xrightarrow{F} & N^{2k} \\ \downarrow & & \downarrow g \\ \bar{\mathcal{C}}_\mu^{k+1} & \xrightarrow{\bar{F}} & Y \end{array}$$

such that

7.5 (a) a framed immersion $\mathcal{C}_\mu^k \to N$ furnished by (4B.23) and (7.4) is regularly homotopic to an imbedding; one such imbedding is chosen and is also denoted $F\colon \mathcal{C}_\mu^k \to N$;

(b) $F_*\colon H_{k-1}(\tilde{\mathcal{C}}_\mu^k) \to K_{k-1}N$ is surjective.

By (3.17), (4A.21) and (4A.36) the data of (7.4) and (7.5) exist for any given local surgery problem, where $\mu = tI_n$, n and t sufficiently large.

Let U denote the regular neighborhood of $F(\mathcal{C}_\mu^k)$ in N^{2k}. Using Diagram (7.4), the fact that $\pi_1\bar{\mathcal{C}}_\mu^{k+1}$ is a free group ((4B.25)(i)) and [W2, Lemma 2.8], we may find a handle-body H having handles of index 0 and 1, such that $Y \cong H \cup_{\partial H} Y'$ and (up to homotopy) H is the regular neighborhood of $\bar{F}(\bar{\mathcal{C}}_\mu^{k+1})$, where we take $\bar{F}$ to be an imbedding ($\bar{\mathcal{C}}_\mu^{k+1}$ has the homotopy type of a bouquet of circles (4B.25)). By obstruction theory and the fact that $\pi_1\mathcal{C}_\mu^k \xrightarrow{\cong} \pi_1\bar{\mathcal{C}}_\mu^{k+1}$, g is homotopic to a map (again denoted g) of quadruples

7.6 $$g: (N;N-\dot{U},U;\partial U) \longrightarrow (Y;Y-\dot{H},H;\partial H)$$

such that $(g|U)_{\#}: \pi_1 U \overset{\cong}{\to} \pi_1 H$. Observe that by the van Kampen Theorem and the fact that $(g|\partial U)_{\#}: \pi_1 \partial U \overset{\cong}{\to} \pi_1 \partial H$, we obtain $(g|N-\dot{U})_{\#}: \pi_1(N-\dot{U}) \overset{\cong}{\to} \pi_1(Y-\dot{H})$. By (7.6) it makes sense to consider the following braid of $\mathbb{Z}\pi$-modules and -homomorphisms which is easily verified to contain the only non-zero kernels of g, $g|U$, $g|\partial U$ and $g|N_0$, where $N_0 = N - \dot{U}$ and s is described below.

7.7
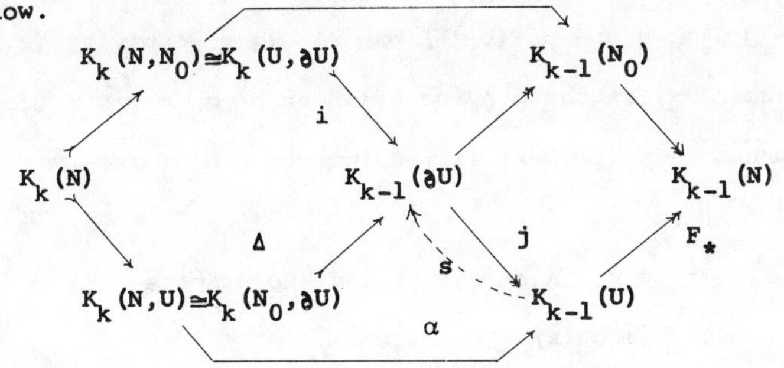

All modules in (7.7) are torsion modules and by (2.8) all, except possibly $K_k(N)$ and $K_{k-1}(N)$, are in $\mathfrak{D}_F^1$. Since $gF: \mathcal{C}_\mu^k \to Y$ extends over $\mathcal{C}_\mu^{k+1}$ (7.4) we have $K_{k-1}(U) = H_{k-1}(U)$, $K_{k-1}(\partial U) = H_{k-1}(\partial U)$ and $K_k(U,\partial U) = H_k(U,\partial U)$. Thus by (4B.40),

7.8 $$K_{k-1}(\partial U) \cong i(K_k(U,\partial U)) + s(K_{k-1}(U))$$

where s in (7.8) is induced by pushing $F\mathcal{C}_\mu^k$ out to ∂U by the first vector in the complementary framing; further, by (4B.45) the $(-1)^k$-form on ∂U, $(K_{k-1}(\partial U),\varphi,\psi)$, is hyperbolic where the summands in (7.8) are totally isotropic and $\varphi|i(K_k(U,\partial U)) \times s(K_{k-1}(U))$ becomes the natural form when $K_{k-1}(U)^\wedge$ is identified with $K_k(U,\partial U)$ by duality in $(U,\partial U)$. Using (7.8) we write $\Delta = (\gamma,\alpha)$ (in 7.7) where $\gamma: K_k(N_0,\partial U) \to K_k(U,\partial U)$ and $\alpha: K_k(N_0,\partial U) \to K_{k-1}(U)$. By (5.18) $\text{im}(\Delta)$ is a subkernel, so if we can construct a $(-1)^{k+1}$-linking form $(K_k(N_0,\partial U),\zeta,\xi)$ where $\zeta(x,y) = \varphi(\gamma(x),\alpha(y))$ $(x,y \in K_k(N_0,\partial U))$ then by definition (2.34) we will have constructed a $(-1)^k$-formation $(K_k(N_0,\partial U),K_{k-1}(U),\Delta,\xi)$.

7.9. To construct ξ, fix copies of $\mathcal{C}_\mu^k$ and $\mathcal{C}_{\bar\mu}^k$ in ∂U, where $\mathcal{C}_{\bar\mu}^k$ is the dual conglomerate (4B.31). Let $x \in K_k(N_0,\partial U)$ and let $\Delta x = [\gamma_x] + [\alpha_x]$, where γ_x = (the (k-1)-chain) $\sum_{i=1}^n (b_i\mathcal{C}_{\bar\mu}^k)\gamma_i$, $\alpha_x = \sum_{j=1}^n (b_j\mathcal{C}_\mu^k)\alpha_j$, $\gamma_i = \sum_g \gamma_{g,i} g \in \mathbb{Z}\pi$, $\alpha_j = \sum_h \alpha_{h,j} h \in \mathbb{Z}\pi$, $\gamma_{g,i}$ and $\alpha_{h,j} \in \mathbb{Z}$, and $b_i\mathcal{C}_{\bar\mu}^k$, $b_j\mathcal{C}_\mu^k$ denote components of the bockstein (4B.3). If $p\colon \tilde N \to N$ is the universal covering, then $\tilde N_0 = p^{-1}N_0$ is simply connected and the (k-1)-chain $\tilde\gamma_x + \tilde\alpha_x = \sum_{g,i} \gamma_{g,i}(S^*_{e,i}\cdot g) + \sum_{h,j} \alpha_{h,j}(S_{e,j}\cdot h)$ is a boundary in $\tilde N_0$, where $p^{-1}(b_j\mathcal{C}_\mu^k) = \bigcup_h S_{e,j}\cdot h$ and $p^{-1}(b_i\mathcal{C}_{\bar\mu}^k) = \bigcup_g S^*_{e,i}\cdot g$. Applying (3.10) in $(M_{\tilde g_0},\tilde N_0)$ (where $M_{\tilde g_0}$ is the mapping cylinder of $\tilde g_0 = \tilde g|\tilde N_0$) together with (4A.25), furnishes a (proper) framed immersion $h\colon (\mathfrak{M}(x), b\mathfrak{M}(x)) \to (N_0,\partial U)$ such that

a) $\mathfrak{M}(x)$ is an identified Moore space,

7.10 b) $[h(b\mathfrak{M}(x))] = \Delta x$, and

c) $[h\mathfrak{M}(x)] = x$ (because Δ is injective).

Recall (4A.28) that $h\mathfrak{M}(x) \pitchfork h\mathfrak{M}(x) \in \mathbb{Z}\pi/J_{-\epsilon}$ denotes the self-intersection number of $h|\mathring{\mathfrak{M}}(x)$, $\epsilon = (-1)^k$. By (4B.35) there are $t \in \mathbb{Z}$ and $\nu \in M_n(\mathbb{Z}\pi,\mathbb{Q}\pi)^\times$ such that $\partial(\sum_{j,i} \mathfrak{M}(j)\nu_{ji}\alpha_i) = t\alpha_x$, where $\mathcal{C}_\mu^k = \bigcup_{i=1}^n \mathfrak{M}(i)$. Let $\beta_x = \sum_{j,i} \mathfrak{M}(j)\nu_{ji}\alpha_i$.

7.11 <u>Definition</u>. $\xi(x) = \frac{1}{t}(\gamma_x \pitchfork \beta_x) - h\mathfrak{M}(x) \pitchfork h\mathfrak{M}(x)$.

$\xi(x)$ is an element of $\mathbb{Q}\pi/J_{-\epsilon}$, $\epsilon = (-1)^k$; since $\frac{1}{t}(\gamma_x \cap \beta_x) \equiv \zeta(x) \pmod{\mathbb{Z}\pi}$ and $h\mathfrak{M}(x) \pitchfork h\mathfrak{M}(x) \in \mathbb{Z}\pi/J_{-\epsilon} \subseteq \mathbb{Q}\pi/J_{-\epsilon}$, we have $\xi(x) \equiv \zeta(x,x) \pmod{\mathbb{Z}\pi}$ (condition (2.5d)(ii)).

Let us show that $\xi(x)$ is independent of the choices made in its definition. By [W1,p. 250] $\frac{1}{t}(\gamma_x \pitchfork \beta_x) \in \mathbb{Q}\pi$ is independent of t; observe also that t and μ determine ν. If $[\alpha_x] = [\sum_i (b_i\mathcal{C}_\mu^k)\alpha_i']$ then by (4B.9) there exist $\lambda_1,\ldots,\lambda_n \in \mathbb{Z}\pi$ such that

$$\partial(\sum_j \mathfrak{M}(j)\lambda_j) = \sum_i (b_i\mathcal{C}_\mu^k)(\alpha_i' - \alpha_i).$$

Setting $\alpha'_x = \sum_i (b_i\mathcal{C}^k_\mu)\alpha'_i$, $\beta'_x = \sum_{i,j} \mathfrak{M}(j)\nu_{ji}\alpha'_i$, and $A = \sum_j \mathfrak{M}(j)\lambda_j$, a calculation ([W1,p.250]) shows

$$7.12 \qquad \frac{1}{t}(\gamma_x \pitchfork \beta_x) - \frac{1}{t}(\gamma_x \pitchfork \beta'_x) = \gamma_x \pitchfork A.$$

On the other hand, if we connect up by tubes $T \approx S^{k-1} \times I$ the copies of $\mathfrak{M}(j)$ (for all $j = 1,\ldots,n$) appearing in the expression $A = \sum_{j=1}^n \mathfrak{M}(j)\lambda_j$, getting the image $\mathfrak{M}_A$ of an identified Moore space in ∂U; then push $\dot{\mathfrak{M}}_A$ into N_0 (fixing $b\mathfrak{M}_A$) getting $\mathfrak{M}'_A$; and finally connect up $\mathfrak{M}'_A$ to $h\mathfrak{M}(x)$ by a tube in $\dot{N}_0$, we obtain the image of a proper framed immersion $h'\colon (\mathfrak{M}'(x), b\mathfrak{M}'(x)) \to (N_0, \partial U)$ satisfying the conditions of (7.10) where $h'(b\mathfrak{M}'(x)) = \sum_i (b_i\mathcal{C}^k_\mu)\gamma_i + \sum_j (b_j\mathcal{C}^k_\mu)\alpha'_j$. Pushing $\mathfrak{M}_A$ into N creates the intersection

$$7.13 \qquad h\dot{\mathfrak{M}}(x) \pitchfork \mathfrak{M}'_A = \gamma_x \pitchfork A.$$

Since $\mathfrak{M}_A$ is a union of disjoint framed imbedded chains $\mathfrak{M}(j)$ ($\mathcal{C}^k_\mu = \cup_j \mathfrak{M}(j)$ is framed imbedded), we may assume that the self-intersection

$$7.14 \qquad \mathfrak{M}'_A \pitchfork \mathfrak{M}'_A = 0.$$

(7.12), (7.13), (7.14) and the construction of h' show that $\xi(x)$ is independent of the choice of cycle representative for $[\alpha_x]$, as long as this representative is a linear combination of the $b_i\mathcal{C}^k_\mu$; this latter condition we now require. A similar argument treats the choice of γ_x.

The remaining ambiguities are the choice of homotopy class of h satisfying (7.10)(c) and then a choice for h within the given regular homotopy class. The latter choice as usual changes $h\mathfrak{M}(x) \pitchfork h\mathfrak{M}(x)$ by an element of J_ϵ, $\epsilon = (-1)^{k+1}$; the former changes $h\mathfrak{M}(x)$ to $h\mathfrak{M}(x) \# f(S^k)$ where $f\colon S^k \to N_0$ is a framed immersion representing an element in the image of $\partial\colon \pi_{k+1}(g_0) \to \pi_k(N_0)$, $g_0 = g|N_0$, and "$\#$" means interior connected sum. (This can be proved as follows: given $h, h'\colon \mathfrak{M}(x) \to N_0$ with $h|b\mathfrak{M}(x) = h'|b\mathfrak{M}(x)$, push them apart at their bocksteins (4A.17) getting $\hat{h},\hat{h}': \hat{\mathfrak{M}}(x) \to N_0$ (where $\hat{\mathfrak{M}}(x)$ is a punctured sphere), form the union $\hat{h}\hat{\mathfrak{M}}(x) \cup \hat{h}'\hat{\mathfrak{M}}(x)$ along common boundary components, and convert it to a k-sphere by doing ambient surgery on its fundamental group; this k-sphere is $f(S^k)$.) Since $K_k(N_0) = 0$, $[f(S^k)] = 0$, whence $f(S^k) \pitchfork f(S^k) = 0 = f(S^k) \pitchfork h\mathfrak{M}(x)$ (the latter intersection being given by the pairing $K_k(N_0) \times K_k(N_0,\partial U) \to \mathbb{Z}\pi$). Hence $\xi: K_k(N_0,\partial U) \to \mathbb{Q}\pi/J_\epsilon$ is well-defined, $\epsilon = (-1)^{k+1}$. The verification that $(K_k(N_0,\partial U),\zeta,\xi)$ is a $(-1)^{k+1}$-linking form is routine and left to the reader. Hence $(K_k(N_0,\partial U),K_{k-1}(U),\Delta,\xi)$ is a $(-1)^k$-formation.

7.15 Definition. $\theta(g,b;\underline{\underline{F}},\mu) = (K_k(N_0,\partial U),K_{k-1}(U),\Delta,\xi)$.

The main result of Chapter 4C is (4C.19) which states that certain data (4C.14) remain unchanged under passage between conglomerates and their compositions. In our setting we may replace homology groups in (4C.14) by kernel groups K_i; we leave to the reader the task of defining $\xi: K_k(N_0,\partial V) \to \mathbb{Q}\pi/J_{-\epsilon}$ where V is the regular neighborhood of a composition of conglomerates. Since the ambient surgery done on U (or V) in the proof of (4C.19) (see (4C.15) and (4C.38)) may be translated to modifications of h in (7.10) which do not change $h\mathfrak{M}(x) \pitchfork h\mathfrak{M}(x)$ or $\frac{1}{t}(\gamma_x \pitchfork \beta_x)$ in (7.11) we obtain the following result:

7.16 Proposition. Theorem (4C.19) remains valid if $\xi: K_k(N_0,\partial V) \to \mathbb{Q}\pi/J_{-\epsilon}$, $\epsilon = (-1)^k$ is added to the data of (4C.14) (where kernel groups replace homology groups).

In the next subsections (numbered (7.17), (7.28), and (7.45)) we examine the choices made in the construction of $\theta(g,b;\underline{\underline{F}},\mu)$, simultaneously proving that its class $\theta(g,b) \in L_1^{(-1)^k}(\mathbb{Q}\pi/\mathbb{Z}\pi)$ depends only on the local surgery problem (g,b) and that the first statement of (7.3) holds.

7.17. To construct $\theta(g,b;\underline{\underline{F}},\mu)$ we chose a framed imbedding $F: \mathcal{C}_\mu^k \to N$

determined by $\underline{\underline{F}}$. Let

7.18
$$\underline{\underline{G}} = \begin{array}{ccc} \mathcal{C}^k_\mu & \xrightarrow{\ G\ } & N \\ \big\downarrow & & \big\downarrow g \\ \bar{\mathcal{C}}^{k+1}_\mu & \xrightarrow{\ \bar{G}\ } & Y \end{array}$$

be a commutative diagram homotopic to $\underline{\underline{F}}$ where G is a framed imbedding. By (4B.23) there is a regular homotopy

7.19. $H\colon \mathcal{C}^k_\mu \times I \to N \times I$ where $F = H|\mathcal{C}^k_\mu \times \{0\}$ and $G = H|\mathcal{C}^k_\mu \times \{1\}$. We may assume $H|b\mathcal{C}^k_\mu \times I$ is the product imbedding. In general position H has at most 1-dimensional self-intersections so

7.20
$$K_k(H(\mathcal{C}^k_\mu \times I), H(\mathcal{C}^k_\mu \times \partial I)) = K_{k-1}(U)$$

By definition H extends to an immersion $U \times I \to N \times I$, also denoted by H. Taking (7.20) as an identification, consider the $(-1)^{k+1}$-linking form $(K_{k-1}(U), \rho, \sigma)$ where

$$\rho\colon K_k(H(U\times I), H(U\times\partial I)) \times K_k(H(U\times I), H(U\times\partial I)) \longrightarrow \mathbb{Q}\pi/\mathbb{Z}\pi$$

7.21 and

$$\sigma\colon K_k(H(U\times I), H(U\times\partial I) \longrightarrow \mathbb{Q}\pi/J_{-\epsilon}, \qquad \epsilon = (-1)^k$$

are the linking and self-linking forms in the $(2k+1)$-manifold $H(U\times I)$. By (4B.33) there is an $(n\times n)$-matrix τ (with entries in $\mathbb{Q}\pi$) such that

7.22
$$H'(b_i\mathcal{C}^k_\mu \times I) \pitchfork H(\mathfrak{M}(j) \times I) = (\tau\mu)_{ij}$$

and if $E \overset{\mu}{\rightarrowtail} F \overset{j}{\twoheadrightarrow} K_{k-1}(U)$ is the resolution corresponding to $\mathcal{C}^k_\mu$ then

7.23 (a)
$$\begin{array}{ccc} F & \xrightarrow{\ \overline{(\tau\mu)}\ } & \bar{E} \\ \big\downarrow j & & \big\downarrow \tilde{j} \\ K_k(U) & \xrightarrow{\ Ad(\rho)\ } & K_k(U)^{\wedge} \end{array}$$

commutes; and 7.23 (b) $\sigma[b_i\mathcal{C}^k_\mu] \equiv \tau_{ii} \pmod{J_{-\epsilon}}$

$\epsilon = (-1)^k$.

Recall the definition (2.36)(a) of $\chi_{(K_{k-1}(U),\rho,\sigma)}\theta$ where θ is a $(-1)^k$-formation.

7.24 <u>Proposition</u>. (a) $\chi_{(K_{k-1}(U),\rho,\sigma)}\theta(g,b;\underline{\underline{F}},\mu) = \theta(g,b;\underline{\underline{G}},\mu)$.

(b) If $(K_{k-1}(U),\psi,\kappa)$ is a $(-1)^{k+1}$-linking form, there exists

$$\underline{\underline{G}} = \begin{array}{ccc} \mathcal{C}^k_\mu & \xrightarrow{G} & N \\ \big\downarrow & & \big\downarrow g \\ \mathcal{C}^{-k+1}_\mu & \xrightarrow{\bar{G}} & Y \end{array} \simeq \underline{\underline{F}}$$

such that G is a framed imbedding and $\theta(g,b;\underline{\underline{G}},\mu) =$
$\chi_{(K_{k-1}(U),\psi,\kappa)}\theta(g,b;\underline{\underline{F}},\mu)$.

<u>Proof</u>. (a) Let $U' = G(U)$ $(= H(U \times \{1\}))$ and let $N_0' = N - \mathring{U}'$. There is an isomorphism $c\colon K_k(N{\times}U) \overset{\cong}{\to} K_k(N,U')$ sending a relative cycle $(e,\partial e) \subseteq (N,U)$ to $(e \cup E,\partial E) \subseteq (N,U')$ where E is ∂e extended along the track of the regular homotopy. Let $\theta(g,b;\underline{\underline{G}},\mu) =$ $(K_k(N_0',U'),K_{k-1}(U'),\Delta',\xi')$ where $\Delta' = (\gamma',\alpha')$. Under the identification c (sometimes made without notational change, $\alpha = \alpha'$ because $\alpha[e,\partial e] = [\partial e]$.

Given $x \in K_k(N_0,\partial U)$, choose $h\colon (\mathfrak{M}(x),b\mathfrak{M}(x)) \to (N_0,\partial U)$ to satisfy (7.10) (a)-(c). Modify h to a framed immersion $f\colon (\eta(x),b\eta(x)) \to (N,U)$, where (in the notation of (7.9))

$$f\eta(x) = h\mathfrak{M}(x) \cup (\bigcup_i (b_i\mathcal{C}^{-k+1}_\mu)_{\gamma_i}). \tag{7.25}$$

This "fills in" the $b\mathcal{C}^k_\mu$-components of the bockstein of $h\mathfrak{M}(x)$ and yields $[f(b\eta(x))] = [\alpha_x]$ in place of (7.10)(b). Observe that since $H|b\mathcal{C}^k_\mu \times I = (F|b\mathcal{C}^k_\mu) \times I$, c sends the relative cycle $(f\eta(x),fb\eta(x))$ to itself.

Assuming $H(\mathcal{C}^k_\mu \times I)$ and $f(\eta(x)) \times I$ intersect tranversely in $N \times I$, a simple calculation shows (c.f. [W1,p.251])

7.26 $H(\mathfrak{M}(i)\times I) \pitchfork f(b\eta(x)) \times I = G(\mathfrak{M}(i)) \pitchfork f(\eta(x)) - F(\mathfrak{M}(i)) \pitchfork f(\eta(x))$.

(Here the chain $f(b\eta(x))$ is taken with multiplicity.) By construction of f, $\gamma(x) = [\gamma_x]$ (resp., $\gamma'(x) = [\gamma'_x]$) where $\gamma_x = \sum_i b_i\mathcal{C}^k_{\bar{\mu}}(F(\mathfrak{M}(i)) \pitchfork f(\eta(x)))$ (resp. $\gamma'_x = \sum_i b_i\mathcal{C}^k_{\bar{\mu}}(G(\mathfrak{M}(i)) \pitchfork f(\eta(x)))$). Hence by (7.26),

7.27 $$\gamma'_x - \gamma_x = \sum_i b_i\mathcal{C}^k_{\bar{\mu}}(H(\mathfrak{M}(i) \times I) \pitchfork f(b\eta(x)) \times I.$$

Setting $\alpha(x) = [\alpha_x]$ where $\alpha_x = \Sigma(b_j\mathcal{C}^k_\mu)\alpha_j$, which by construction (7.25) equals $f(b\eta(x))$,

$$\begin{aligned}\gamma'(x) - \gamma(x) &= \sum_{i,j} [b_i\mathcal{C}^k_{\bar{\mu}}](H(\mathfrak{M}(i)\times I) \pitchfork H'(b_j\mathcal{C}^k_\mu\times I))\alpha_j \quad \text{(by (7.27))}\\ &= \sum_{i,j} [b_i\mathcal{C}^k_{\bar{\mu}}](\overline{\tau\mu})_{ij}\alpha_j) \quad \text{(by (7.22))}\\ &= \sum_j (\mathrm{Ad}(\rho)[b_j\mathcal{C}^k_\mu])\alpha_j \quad \text{(by (7.23)(a) and (4B.44))}\\ &= \mathrm{Ad}(\rho)(\alpha(x)).\end{aligned}$$

To complete the proof of (7.24)(a), it remains to show that $\xi'(x) = \xi(x) + \sigma(\alpha(x))$. Let $x \in K_k(N_0,\partial U)$ be represented by a framed immersion h as in (7.10). The identification of $K_k(N_0,\partial U)$ with $K_k(N'_0,\partial U')$ converts h to h' representing $x \in K_k(N'_0,\partial U')$ by appending to $h(b\mathfrak{M}(x))$ the track of the regular homotopy H (compare the definition of c above). Since we have assumed H fixes $b\mathcal{C}^k_\mu$, $h\mathfrak{M}(x) \pitchfork h\mathfrak{M}(x) = h'\mathfrak{M}(x) \pitchfork h'\mathfrak{M}(x)$ in (7.11); hence $\xi'(x) - \xi(x)$ is carried in the $\frac{1}{t}(\alpha_x \pitchfork \beta_x)$-component of $\xi(x)$. This is easily shown to be $\sigma(\alpha(x))$ using (7.23)(b), (7.22) and (7.27). Details are left to the reader.

Conversely, given a $(-1)^{k+1}$-linking form $(K_{k-1}(U),\psi,\varkappa)$, there is by (6.16) a regular homotopy $H\colon \mathcal{C}^k_\mu \times I \to N \times I$ of F to an imbedding $G(= H|\mathcal{C}^k_\mu \times \{1\})$ such that the functions of (7.21) are the given ψ and $\varkappa$.

By (7.24)(a), (7.24)(b) is proved.

7.28. Next we relate composition of conglomerates (4C.2) to stabilization of a formation (2.36)(c). The important algebraic point is (2.45): all short exact sequences in $\mathfrak{D}_F^1$ arise as short exact sequences of cokernels. To use this result, given $\underline{\underline{F}}$ in (7.4), observe that changing $F(\mathcal{C}_\mu^k) \subseteq N$ by adding to it m disjoint framed imbedded k-discs $B_1^k,\ldots,B_m^k$ (where $\partial B_1^k,\ldots,\partial B_m^k$ are regarded as bockstein components) yields a conglomerate constructed according to $\mu \perp I_m$ (see (2.44)). $\underline{\underline{F}}$ is altered in the obvious way and $\theta(g,b;\underline{\underline{F}},\mu)$ is unchanged. This together with (2.45) proves the following result, where $E(\nu,\mu) = (\text{cok } \nu \rightarrowtail \text{cok } \mu\nu \twoheadrightarrow \text{cok } \mu)$.

7.29 <u>Lemma</u>. (a) Given any short exact sequence $E = (J \rightarrowtail H \twoheadrightarrow K_{k-1}(F\mathcal{C}_\mu^k))$ in $\mathfrak{D}_F^1$, we may assume there is $\nu \in M_n(\mathbb{Z}\pi,\mathbb{Q}\pi)^\times$ such that $E \cong E(\nu,\mu)$ (rel. $K_{k-1}(F\mathcal{C}_\mu^k)$).

(b) Given $\underline{\underline{F}}$ in (7.4), with η in place of μ, and any short exact sequence $E = (J \rightarrowtail K_{k-1}(F\mathcal{C}_\eta^k) \twoheadrightarrow H)$ in $\mathfrak{D}_F^1$, we may assume there exist $\nu,\mu \in M_n(\mathbb{Z}\pi,\mathbb{Q}\pi)^\times$ such that $\eta = \mu\nu$ and $E \cong E(\nu,\mu)$ (rel. $K_{k-1}(F\mathcal{C}_\mu^k)$).

On the geometric side, given $\nu \in M_n(\mathbb{Z}\pi,\mathbb{Q}\pi)^\times$ and $\underline{\underline{F}}$ in (7.4) satisfying (7.5)(a)-(b), replace $F\mathcal{C}_\mu^k \subseteq N$ by the composition $F(\mathcal{C}_\mu^k)\mathcal{C}_\nu^k$ where $\mathcal{C}_\nu^k \subseteq N$ is trivially imbedded (4B.58) (compare (4C.3) and (7.36)), and convert $F(\mathcal{C}_\mu^k)\mathcal{C}_\nu^k$ to a copy of $\mathcal{C}_{\mu\nu}^k$ in N using (4C.19). According to (4C.19) the latter is the image of a framed imbedding, denoted $\sigma_\nu F\colon \mathcal{C}_{\mu\nu}^k \to N$, and well defined up to isotopy; further, $g(\sigma_\nu F)\colon \mathcal{C}_{\mu\nu}^k \to Y$ extends to $\sigma_\nu\bar{F}\colon \bar{\mathcal{C}}_{\mu\nu}^{k+1} \to Y$ in the obvious way ($\mathcal{C}_\nu^k$ is trivially imbedded).

7.30 <u>Definition</u>. $$\sigma_\nu\underline{\underline{F}} = \begin{array}{ccc} \mathcal{C}_{\mu\nu}^k & \xrightarrow{\sigma_\nu F} & N \\ \downarrow & & \downarrow g \\ \bar{\mathcal{C}}_{\mu\nu}^{k+1} & \xrightarrow{\sigma_\nu \bar{F}} & Y \end{array}$$ is the <u>stabilization of $\underline{\underline{F}}$ by</u> ν.

Recall from (2.36)(c) the definition of $\sigma_E\theta$ where E is a short exact sequence in $\mathfrak{D}_F^1$ and θ is a formation.

7.31 <u>Proposition</u>. (a) $\sigma_{E(\nu,\mu)}\theta(g,b;\underline{\underline{F}},\mu) = \theta(g,b;\sigma_\nu\underline{\underline{F}},\mu\nu)$.

(b) Given $\underline{\underline{F}}$ satisfying (7.5) and a short exact sequence $E = (J \rightarrowtail H \twoheadrightarrow K_{k-1}(F\mathcal{C}_\mu^k))$ in $\mathfrak{D}_F^1$, there is $\nu \in M_n(\mathbb{Z}\pi,\mathbb{Q}\pi)^\times$ such that $E \cong E(\nu,\mu)$ and $\theta(g,b;\sigma_\nu\underline{\underline{F}},\mu\nu) = \sigma_E\theta(g,b;\underline{\underline{F}},\mu)$.

(c) Let $\underline{\underline{G}} = \begin{array}{ccc} \mathcal{C}_\eta^k & \longrightarrow & N \\ \downarrow & & \downarrow g \\ \bar{\mathcal{C}}_\eta^{k+1} & \longrightarrow & Y \end{array}$ satisfy the conditions of (7.5) and let $\theta(g,b;\underline{\underline{G}},\eta) = \sigma_E\theta$ where $E = (J \rightarrowtail H_{k-1}(\tilde{\mathcal{C}}_\eta^k) \twoheadrightarrow H)$ is a short exact sequence of elements of $\mathfrak{D}_F^1$ and θ is a $(-1)^k$-formation. Then $\eta = \mu\nu$ for some $\mu,\nu \in M_n(\mathbb{Z}\pi,\mathbb{Q}\pi)^\times$, and there exists $\underline{\underline{F}} = \begin{array}{ccc} \mathcal{C}_\mu^k & \xrightarrow{F} & N \\ \downarrow & & \downarrow \\ \bar{\mathcal{C}}_\mu^{k+1} & \xrightarrow{\bar{F}} & Y \end{array}$ satisfying (7.5) such that $\theta(g,b;\underline{\underline{F}},\mu) = \theta$.

<u>Proof</u>. (a) Theorem (4C.19) and (7.16) allow us to replace $\sigma_\nu F$ by the framed imbedding canonically associated to it, $\mathcal{C}_\mu^k\mathcal{C}_\nu^k \to N$, also denoted $\sigma_\nu F$. We will use the notation of (4C.2) and (4C.3); let E_i (resp. D_j) denote the image of $E_{e,i}$ (resp. $D_{e,j}$) under the canonical projection $\tilde{\mathcal{C}}_\mu^k \to \mathcal{C}_\mu^k$ (resp. $\tilde{\mathcal{C}}_\nu^k \to \mathcal{C}_\nu^k$), $i = 1,\ldots,n$ $(j = 1,\ldots,n)$. The notation often replaces the image of an imbedding into N by its domain.

Let U_μ (resp. $U_\nu, U_{\mu,\nu}$) be the regular neighborhood in N of $\mathcal{C}_\mu^k$ (resp. $\mathcal{C}_\nu^k, \mathcal{C}_\mu^k\mathcal{C}_\nu^k$), and let $C^{2k} \subseteq N$ be a 2k-disc such that $C^{2k} \cap \mathcal{C}_\mu^k = \cup E_i$ and $bU_\nu = C^{2k} \cap U_{\mu,\nu}$ is a regular neighborhood of $b\mathcal{C}_\nu^k$ in N. Let the map $c'\colon (N,U_{\mu,\nu}) \to (N,U_\mu \vee (U_\nu/bU_\nu))$ collapse C^{2k} to a point; strictly speaking the target of c' should have a regular neighborhood of $\mathcal{C}_\mu^k/\cup E_i$ in place of U_μ, but

$$K_j(U_\mu) = K_j(\mathcal{C}_\mu^k) \cong K_j(\mathcal{C}_\mu^k/\cup E_i),\ j \neq 0,1, \tag{7.32}$$

so we ignore this point.

Consider the map of exact sequences induced by the inclusion $(N,U_\mu) \hookrightarrow (N,U_\mu \vee (U_\nu/bU_\nu))$

7.33 <u>Figure</u>.

$$\begin{array}{ccccccc}
K_k(N) \rightarrowtail K_k(N,U_\mu) & \xrightarrow{\alpha} & K_{k-1}(U_\mu) & \longrightarrow\!\!\!\!\rightarrow & K_{k-1}(N) \\
\downarrow = \qquad \downarrow i'_* & & \downarrow i_* & & \downarrow = \\
K_k(U_\mu \vee (U_\nu/bU_\nu)) \rightarrow K_k(N) \xrightarrow{\ell} K_k(N,U_\mu \vee (U_\nu/bU_\nu)) & \xrightarrow{\alpha''} & K_{k-1}(U_\mu \vee (U_\nu/bU_\nu)) & \rightarrow & K_{k-1}(N)
\end{array}$$

Since $\mathcal{C}^k_\nu$ is trivially imbedded in N and $K_k(U_\mu) = 0$, ℓ is injective. Since $K_{k-1}(\mathcal{C}^k_\nu/b\mathcal{C}^k_\nu) = 0$ (4B.5), i'_* is surjective; it is injective because $\mathcal{C}^k_\nu$ is trivially imbedded. Thus i'_* and i_* are isomorphisms. In the commutative diagram (where $c = c'|U_{\mu,\nu}$)

7.34 <u>Figure</u>.

$$\begin{array}{ccccccc}
K_{k-1}(\mathcal{C}^k_\nu) & \rightarrowtail & K_{k-1}(\mathcal{C}^k_\mu\mathcal{C}^k_\nu) & & \longrightarrow\!\!\!\!\rightarrow & & K_{k-1}(\mathcal{C}^k_\mu\mathcal{C}^k_\nu/\mathcal{C}^k_\nu) \\
\cong\downarrow a & & \downarrow\cong & & & & \cong\downarrow b \\
\ker(c_*) & \rightarrowtail & K_{k-1}(U_{\mu,\nu}) & \xrightarrow{c_*} & K_{k-1}(U_\mu \vee (U_\nu/bU_\nu)) & \xleftarrow[\cong]{i_*} & K_{k-1}(U_\mu)
\end{array}$$

we use

7.35 $$\mathcal{C}^k_\mu\mathcal{C}^k_\nu/\mathcal{C}^k_\nu = \mathcal{C}^k_\mu/\cup E_i$$

and (7.32) to define b, and (4B.4)(ii) to show the top sequence is short exact; a is the induced isomorphism. Using (7.32) and (7.35) to identify $K_{k-1}(\mathcal{C}^k_\mu\mathcal{C}^k_\nu/\mathcal{C}^k_\nu)$ with $K_{k-1}(\mathcal{C}^k_\mu)$, together with (4C.3), (4C.4), and (4C.5), the following is easily verified.

7.36 $$E(\nu,\mu) = (K_{k-1}(\mathcal{C}^k_\nu) \rightarrowtail K_{k-1}(\mathcal{C}^k_\mu\mathcal{C}^k_\nu) \twoheadrightarrow K_{k-1}(\mathcal{C}^k_\mu)).$$

To verify (2.41)(a) we show c_* and c'_* are surjective and $\alpha'|\ker c'_*$ is an isomorphism in

$$\begin{array}{ccccccc}
\ker c'_* & \rightarrowtail & K_k(N,U_{\mu,\nu}) & \xrightarrow{c'_*} & K_k(N,U_\mu \vee (U_\nu/bU_\nu)) & \xleftarrow[\cong]{i'_*} & K_k(N,U_\mu) \\
\downarrow \alpha'|\ker c'_* & & \downarrow \alpha' & & \downarrow \alpha'' & & \downarrow \alpha \\
\ker c_* & \longrightarrow & K_{k-1}(U_{\mu,\nu}) & \longrightarrow & K_{k-1}(U_\mu \vee (U_\nu/bU_\nu)) & \xleftarrow[\cong]{i_*} & K_{k-1}(U_\mu)
\end{array}$$

Surjectivity of c_* follows from (7.34); that of c'_* now follows from (7.37) and the isomorphisms cok $\alpha' \xrightarrow{\cong}$ cok α'' (induced by c_*--both cokernels are $K_{k-1}(N)$), and $(c'_*|\ker \alpha')$: $\ker \alpha' \xrightarrow{\cong} \ker \alpha''$ (each kernel is $K_k(N)$). That $\alpha'|\ker c'_*$ is an isomorphism follows from the preceding isomorphisms and (7.37).

To verify (2.41)(b) we show that $\gamma' = p^{\wedge}\cdot\gamma\cdot q$ where $p = (i_*)^{-1}c_*$, $q = (i'_*)^{-1}c'_*$ (see (7.37)). To do this it suffices to show that

7.38 $$\varphi'(\gamma'(x),y) = \varphi(\gamma\cdot q(x),p(y))$$

for all $x \in K_k(N,U_{\mu,\nu})$ and for $y = [b_1\mathcal{C}_\mu^k],\ldots,[b_n\mathcal{C}_\mu^k]$, $[b_1\mathcal{C}_\nu^k],\ldots,[b_n\mathcal{C}_\nu^k]$ $\in K_{k-1}(U_{\mu,\nu})$, where $\varphi'\colon K_k(U_{\mu,\nu},\partial U_{\mu,\nu}) \times K_{k-1}(U_{\mu,\nu}) \to \mathbb{Q}\pi/\mathbb{Z}\pi$ and $\varphi\colon K_k(U_\mu,\partial U_\mu) \times K_{k-1}(U_\mu) \to \mathbb{Q}\pi/\mathbb{Z}\pi$ are linking forms given by duality (compare (2.35)(a)). If $y = [b_i\mathcal{C}_\nu^k]$ the right side of (7.38) vanishes because $c_*[b_i\mathcal{C}_\nu^k] = 0$; $\varphi'(\gamma'(x),[b_i\mathcal{C}_\nu^k]) = \frac{1}{t}(\eta \pitchfork \omega)$ where η is a cycle representing $\gamma'(x)$ and ω is a k-chain carried by $\mathcal{C}_\nu^k$ such that $\partial\omega = t(b_i\mathcal{C}_\nu^k)$. Since $\mathcal{C}_\nu^k$ is trivially imbedded, $\eta \pitchfork \omega = 0$. Next suppose $y = [b_j\mathcal{C}_\mu^k]$. Then there exist (by (4B.35)) $\sigma_i \in \mathbb{Z}\pi$ and $t \in \mathbb{Z}$ such that

7.39 $$t(b_j\mathcal{C}_\mu^k) = \partial(\sum_{i=1}^{n} \mathfrak{M}(i)\sigma_i).$$

Similarly there is $s \in \mathbb{Z}$ and an $(n\times n)$-matrix λ such that

7.40 $$s(b_i\mathcal{C}_\nu^k) = \partial(\sum_{\ell=1}^{n} \mathcal{N}(\ell)\lambda_{\ell i}), \qquad i = 1,\ldots,n.$$

Multiplying (7.39) by s if necessary, we may assume $s|\sigma_i$, $i = 1,\ldots,n$. A calculation shows

7.41 $$t(b_j\mathcal{C}_\mu^k) = \partial(\sum_{i=1}^{n} ((\mathfrak{M}(i)-\mathring{E}_i)\sigma_i - \sum_{\ell=1}^{n} \mathcal{N}(\ell)\lambda_{\ell i}\sigma_i s^{-1})).$$

Since $\gamma'(x)$ (resp. $\gamma q(x)$) is essentially given by the intersection of a k-cycle τ with $\mathcal{C}_\mu^k\mathcal{C}_\nu^k$ (resp. $c'(\tau)$ with $\mathcal{C}_\mu^k$) where τ represents $x \in K_k(N,U_{\mu,\nu})$ and we can assume $\tau \cap C^{2k} = \emptyset$, we have

$\varphi(\gamma q(x),p[b_j\mathcal{C}^k_\mu]) = \frac{1}{t}(\chi \pitchfork (\sum_i \mathfrak{m}(i)\sigma_i))$ and

$\varphi'(\gamma'(x),[b_j\mathcal{C}^k_\mu]) = \frac{1}{t}(\chi \pitchfork (\sum_i((\mathfrak{m}(i) - \dot{E}_i)\sigma_i - \sum_\ell \eta(\ell)\lambda_{\ell i}\sigma_i s^{-1}))$ where χ is the relative k-cycle $\tau \cap U_{\mu,\nu}$. Since $\mathcal{C}^k_\nu$ is trivially imbedded we may again assume $\tau \cap \eta(\ell) = \emptyset$, for all ℓ, so (7.38) is true in case $y = [b_j\mathcal{C}^k_\mu]$. Thus (2.41)(b) is satisfied.

To verify (2.41)(c), observe first that ker $c'_* =$ ker q is generated by the classes $[E_i] \in K_k(N,U_{\mu,\nu})$; this follows from the fact that $\{[\partial E_i] = [b_i\mathcal{C}^k_\nu] \mid i = 1,\dots,n\}$ generates $K_{k-1}(\mathcal{C}^k_\nu)$, and because a in (7.34) and $\alpha'|\ker c'_*$ in (7.37) are isomorphisms. It follows easily from the construction of $\xi': K_k(N,U_{\mu,\nu}) \to \mathbb{Q}\pi/J_{-\epsilon}$, $\epsilon = (-1)^k$, in (7.9) and the definition of ζ' in the formation $\theta(g,b;\underline{\underline{G}},\mu\nu)$ (see (2.34)) that $\xi'|\ker q \equiv 0 \equiv \zeta'|(\ker q) \times K_k(N,U_{\mu,\nu})$. Thus, $\xi'(x + z) = \xi'(x)$, for all $x \in K_k(N,U_{\mu,\nu})$, $z \in \ker q$. From this it follows easily that $\xi' = \xi\cdot q$, where $\xi: K_k(N,U_\mu) \to \mathbb{Q}\pi/J_{-\epsilon}$ is constructed in (7.9).

(b). This follows immediately from (a) and (7.29)(a).

(c). The geometric ingredient is the following lemma, which requires that $\underline{\underline{G}}$ be stabilized. By (7.29)(b), we may assume $E = E(\nu,\mu)$ where $\eta = \mu\nu$.

7.42 <u>Lemma</u>. There exist $t \in \mathbb{Z}$ and $\underline{\underline{F}}$ in (7.4) satisfying (7.5) such that

$$\sigma_{tI_n}\underline{\underline{G}} = \sigma_{t\nu}\underline{\underline{F}}.$$

Assuming the Lemma, observe that

$$\begin{aligned}\sigma_{E(t\nu,\mu)}\theta &= \sigma_{E(tI_n,\eta)}(\sigma_E\theta), && \text{by (2.47)}\\ &= \sigma_{E(tI_n,\eta)}\theta(g,b;\underline{\underline{G}},\eta)\\ &= \theta(g,b;\sigma_{tI_n}\underline{\underline{G}},\eta), && \text{by part (a)}\\ &= \theta(g,b;\sigma_{t\nu}\underline{\underline{F}},\mu), && \text{by Lemma (7.42)}\end{aligned}$$

$$= \sigma_{E(t\nu,\mu)}\theta(g,b;\underline{\underline{F}},\mu), \quad \text{by part (a).}$$

By (2.43), $\theta = \theta(g,b;\underline{\underline{F}},\mu)$.

Proof (of (7.42). By (4C.19) and (7.16) we may replace $G(\mathcal{C}^k_{\mu\nu}) \subseteq N$ by a framed imbedded copy of $\mathcal{C}^k_\mu\mathcal{C}^k_\nu$. The proof first constructs imbedded k-discs $E_i \subseteq N$ such that $\partial E_i = b_i\mathcal{C}^k_\nu$, $\dot{E}_i \cap \mathcal{C}^k_\mu\mathcal{C}^k_\nu = \emptyset$, and E_i is framed compatibly with $\mathcal{C}^k_\mu\mathcal{C}^k_\nu - \dot{\mathcal{C}}^k_\nu$ (in the obvious sense) $i = 1,\ldots,n$. Second, it shows $\mathcal{C}^k_{t\nu}$ is trivially imbedded for t sufficiently large, where $\mathcal{C}^k_{t\nu}$ arises from $\mathcal{C}^k_\nu\mathcal{C}^k_{tI_n}$ in (4C.19) and $\mathcal{C}^k_{tI_n}$ is trivially imbedded. Setting $F(\mathcal{C}^k_\mu) = (\mathcal{C}^k_\mu\mathcal{C}^k_\nu - \dot{\mathcal{C}}^k_\nu) \cup (\bigcup_i E_i)$ is sufficient to define the required $\underline{\underline{F}}$.

Let $\theta(g,b;\underline{\underline{G}},\eta) = (K_k(N,U),K_{k-1}(U),\Delta,\xi)$, where $\Delta = (\gamma,\alpha)$ as usual (see (7.8)). The class $a_i = \alpha^{-1}([b_i\mathcal{C}^k_\nu]) \in K_k(N,U) = K_k(N_0,\partial U)$ satisfies $\gamma(a_i) = 0$ by (2.41)(b) and chasing in Diagram (2.42). (Applying α^{-1} to $[b_i\mathcal{C}^k_\nu]$ makes sense because $J = K_{k-1}(\mathcal{C}^k_\nu)$ by (7.36) and $\alpha_2|L$ is an isomorphism in (2.41)(a).) Since the composition $K_k(N_0,\partial U) \overset{\Delta}{\rightarrowtail} K_{k-1}(\partial U) \twoheadrightarrow K_{k-1}(N_0)$ is zero and $g|N_0$ is (k-1)-connected, the existence of a_i implies there is a framed immersion $h_i\colon (D^k,S^{k-1}) \to (N_0,\partial U)$ such that $h_i(S^{k-1}) = b_i\mathcal{C}^k_\nu$ and $h_i(\dot{D}^k) \cap \partial U = \emptyset$, $i = 1,\ldots,n$. (In fact, $[h_i(D^k)] = a_i$.) The following lemma shows that the h_i may be deformed to imbeddings relative to their boundaries; we then set $h_i(D^k) = E_i$.

7.43 Lemma. (a) $h_i(D^k) \pitchfork h_j(D^k) = 0 \in \mathbb{Z}\pi$, $i,j = 1,\ldots,n$.

(b) $\xi[h_i(D^k)] = 0 \in \mathbb{Q}\pi/J_{(-1)^{k+1}}$, $i = 1,\ldots,n$.

Proof. Let $h_i(D^k) \pitchfork h_j(D^k) = \lambda_{ij} \in \mathbb{Z}\pi$ (λ_{ii} is the intersection number of h_i with h_i', a copy of h_i pushed a non-zero distance along a vector field from the framing). If λ denotes the corresponding $(n\times n)$-matrix, it suffices to show $\bar{\nu}\lambda\nu = 0_n$, since ν is $\mathbb{Q}\pi$-invertible. Let Σ_i denote the n-cycle $\eta(i) \cup (\bigcup_j h_j(D^k)\nu_{ji})$, where $\mathcal{C}^k_\nu = \cup\,\eta(i)$. Then $[\Sigma_i] = 0$ in $K_k(N)$ because Σ_i can be pushed into N_0 so that the image of

$[\Sigma_i]$ in $K_k(N,N_0) = K_k(U,\partial U)$ is zero. Now the equation

7.44 $$\Sigma_i \pitchfork \Sigma_j = \sum_{m,\ell} \bar{\nu}_{mi}\, \lambda_{m\ell}\, \nu_{\ell j}$$

holds because for each fixed m and ℓ, $h_m(D^k) \pitchfork h_\ell(D^k)$ contributes $(h_m(D^k)\nu_{mi}) \pitchfork (h_\ell(D^k)\nu_{\ell j}) = \bar{\nu}_{mi}\lambda_{m\ell}\nu_{\ell j}$ to $\Sigma_i \pitchfork \Sigma_j$; so summing over m and $\ell = 1,\ldots,n$ yields (7.44). Since $[\Sigma_i] = 0$ for all i and (7.44) is the (i,j)-entry in $\bar{\nu}\lambda\nu$, the latter is the zero matrix. Part (b) follows from (2.43)(c) since (as observed above) $J = K_{k-1}(\mathcal{C}^k_\nu)$ and the classes $[b_1\mathcal{C}^k_\nu],\ldots,[b_n\mathcal{C}^k_\nu]$ generate J.

Using (7.43)(a), mutual intersections among the $h_i(D^k)$ may be removed by the Whitney device. The framed immersion h_i satisfies the conditions of (7.10) where $x = [h_i(D^k)]$. Hence $\gamma_x = 0$ in (7.11) so the remaining term in (7.11) vanishes; the Whitney device now removes self-intersections of $h_i\colon (D^k,S^{k-1}) \to (N_0,\partial U)$, $i = 1,\ldots,n$.

To show that $\mathcal{C}^k_\nu$ is trivially imbedded, it suffices to show that the (singular) sphere Σ_j constructed in the proof of (7.43) bounds a (singular) $(k+1)$-disc Δ_j in N_0; in this case Δ_j is (by construction of Σ_j) the image of a relative identified Moore space $\bar{\mathfrak{m}}(j)$ $(j = 1,\ldots,n)$ in such a way that $\mathcal{C}^k_\nu$ bounds $\bar{\mathcal{C}}^{k+1}_\nu = \bigcup_{j=1}^{n} \bar{\mathfrak{m}}(j)$. But each Σ_j arises from a (possibly non-zero) element $\sigma_j \in \pi_{k+1}(g|N_0)$, so the Δ_j do not in general exist.

By the Serre-Hurewicz Theorem (mod the class of finite abelian groups) there is $t \in \mathbb{Z}$ such that $t\sigma_j = 0$, $j = 1,\ldots,n$. It follows from (4C.16) that composing $\mathcal{C}^k_\nu$ with a trivially embedded $\mathcal{C}^k_{tI_n} = \bigcup_{i=1}^{n} \mathfrak{m}^k_t(i)$ amounts to replacing each σ_j by $t\sigma_j = 0$. This furnishes the Δ_j so the proof of (7.42) is complete.

7.45. The final ingredient in the proof that $\theta(g,b;\underline{\underline{F}},\mu)$ is independent of $\underline{\underline{F}}$ is a geometric realization of a special case of Schanuel's Lemma (see [P3,1.7]).

7.46 <u>Proposition.</u> Let $(g,b)\colon (N,\nu_N) \to (Y,\xi)$ be a highly-connected local surgery problem and let $\underline{\underline{F}} = \begin{array}{ccc} \mathcal{C}^k_{tI_p} & \xrightarrow{F} & N \\ \downarrow p & & \downarrow g \\ \mathcal{C}^{-k+1}_{tI_p} & \xrightarrow{\bar{F}} & Y \end{array}$ satisfy (7.5) where I_p is the $(p\times p)$-identity matrix, $\mathcal{C}^k_{tI_p} = \bigcup_{i=1}^{p} \mathfrak{M}^k_t(i)$, the disjoint union of Moore spaces, $p = m + n$, and $[F(b\mathfrak{M}^k_t(i))] = 0$, $i > n$. If $\{y_1,\ldots,y_m\}$ generates $K_{k-1}(N)$, there is a map $\underline{\underline{G}} = \begin{array}{ccc} \mathcal{C}^k_{tI_p} & \xrightarrow{G} & N \\ \downarrow p & & \downarrow g \\ \mathcal{C}^{-k+1}_{tI_p} & \xrightarrow{\bar{G}} & Y \end{array}$ satisfying (7.5) such that

7.47 $$[G(b\mathfrak{M}^k_t(i))] = \begin{cases} y_i, & i = 1,\ldots,m \\ 0, & i = m+1,\ldots,p = m+n, \end{cases}$$

and $\theta(g,b;\underline{\underline{F}},tI_p) = \theta(g,b;\underline{\underline{G}},tI_p)$ in $F_1^{(-1)^k}(\mathbb{Q}\pi/\mathbb{Z}\pi)$.

<u>Proof</u>. Let $x_i = [F(b\mathfrak{M}^k_t(i))]$, $i = 1,\ldots,n$; then $\{x_1,\ldots,x_n\}$ generates $K_{k-1}(N)$ by (7.5)(b). The following three types of operations suffice to convert the ordered set $\{x_1,\ldots,x_n,\overbrace{0,\ldots,0}^{m \text{ terms}}\}$ to the ordered set $\{y_1,\ldots,y_m,\overbrace{0,\ldots,0}^{n \text{ terms}}\}$:

7.48
a) Convert x_i to $\pm x_i g$ for some $g \in \pi$ and leave x_r unchanged, $r \neq i$.

b) Permute the elements.

c) Convert x_i to $x_i + x_\ell$, and leave x_r unchanged, $r \neq i$.

The geometric realization of these operations will convert $\underline{\underline{F}}$ to $\underline{\underline{G}}$ (satisfying (7.47)) without changing $\mathfrak{g}$ (7.11) or the data of (4C.14).

In (7.48)(a) x_i is changed to $x_i g$ by altering the choice of path from $F(\mathfrak{M}^k_t(i))$ to the basepoint in N (4B.17). To change x_i to $-x_i$ use the orientation-reversing homeomorphism $\rho\colon \mathfrak{M}^k_t(i) \to \mathfrak{M}^k_t(i)$, where ρ is given in terms of the model for $\mathfrak{M}^k_t$ in (3.11) by

7.49 $$\rho(t_1,\ldots,t_{k+1}) = (1-t_1,t_2,\ldots,t_{k+1}).$$

The operation (7.48)(b) is realized by reordering the $\mathfrak{M}_t^k(i)$.

To realize (7.48)(c) we first recall from (3.14) how elements of $\pi_k(N;Z_t)$ are added. Let $j \in \{1,2\}$; a set of t k-discs $D_{j1},\ldots,D_{jt} \subseteq \dot{I}_j^{k+1}$ is chosen and a map $F_j\colon \mathfrak{M}_t^k(j) \to N$ is identified with a map $\widetilde{F}_j\colon (\dot{I}_j^{k+1} - \bigcup_{\ell=1}^{t} \dot{D}_{j\ell}) \to N$ where $\widetilde{F}_j|\partial D_{ji} = \widetilde{F}_j|\partial D_{j\ell}$, for all i and ℓ, and $\widetilde{F}(J \cup (\bigcup_\ell J_\ell)) \subseteq$ "base disc" (see (3.20)). $\widetilde{F_1 * F_2}$ is constructed by taking the union of $\widetilde{F}_1$ and $\widetilde{F}_2$ over $\widetilde{F}_1|\{(1,t_2,\ldots,t_{k+1})\} = \widetilde{F}_2|\{(0,t_2,\ldots,t_{k+1})\}$ and removing the common disc E along which the identification is made. Let $V = (\dot{I}_1^{k+1} - \bigcup_\ell D_{1\ell}) \cup (\dot{I}_2^{k+1} - \bigcup_\ell D_{2\ell})$, where the union is taken along E. Observe that

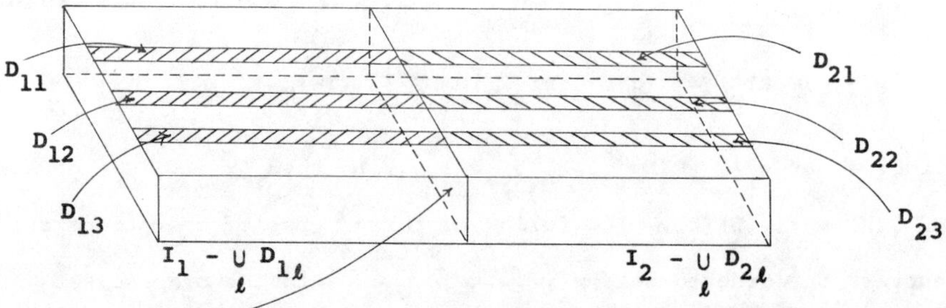

7.50 Figure.

$$E = \{(1,t_2,\ldots,t_{k+1}) \in \dot{I}_1^{k+1}\} = \{(0,t_2,\ldots,t_{k+1}) \in \dot{I}_2^{k+1}\}$$

7.51 $$V/(\partial D_{1i} \approx \partial D_{1\ell};\ \partial D_{2i} \approx \partial D_{2\ell})$$

with the identifications taken for all i and ℓ, is homeomorphic to $\mathfrak{M}_t^k(1) \cup \mathfrak{M}_t^k(2)$, where the union is taken along $\bar{E} = E/((\partial D_{1i} \cap E) \approx (\partial D_{1\ell} \cap E))$, a relative Moore space $\bar{\mathfrak{M}}_t^k$ (see (3.20)).

We will use another model W of V to see operation (7.48)(c). Imbed S^{k-1} in R^k in the standard way and let $v_1,v_2,v_3 \in R$ be the three vertices of an equilateral triangle whose center is at the origin. Consider the space $Z = S^{k-1} * \{v_1,v_2,v_3\} \subseteq R^k \times R^2$ and the homeomorphism $h\colon Z \to Z$ induced by rotating the vertices through an angle of $2\pi/3$. Choose disjoint (k-1)-discs $B_1,\ldots,B_t \subseteq S^{k-1}$ and points $a_1,\ldots,a_t \in Z$ near v_1. Let $b_i = h(a_i)$, $c_i = h(b_i)$.

7.52 Figure.

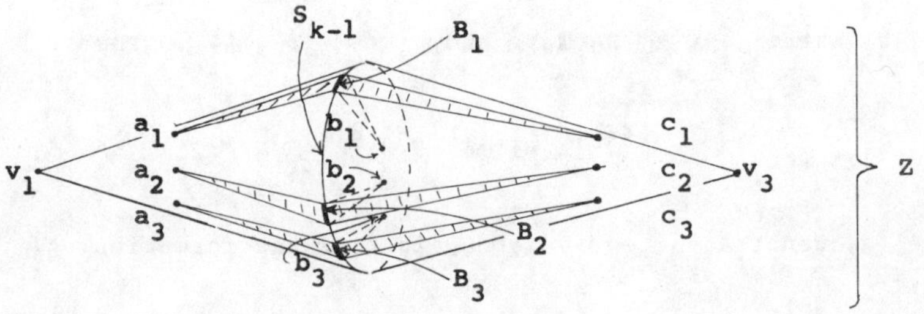

Then $h|B_\ell * a_\ell : B_\ell * a_\ell \xrightarrow{\approx} B_\ell * b_\ell$, $h|B_\ell * b_\ell : B_\ell * b_\ell \xrightarrow{\approx} B_\ell * c_\ell$, and $h|B_\ell * c_\ell : B_\ell * c_\ell \xrightarrow{\approx} B_\ell * a_\ell$, $\ell = 1,\dots,t$. Let $W = Z - \bigcup_\ell (\mathring{B}_\ell * a_\ell \cup \mathring{B}_\ell * b_\ell \cup \mathring{B}_\ell * c_\ell)$. There is a homeomorphism $f\colon V \to W$ where $f(\partial D_{1\ell} - \partial D_{1\ell} \cap E) = (\partial B_\ell) * a_\ell$, $f(\partial D_{1\ell} \cap E) \equiv f(\partial D_{2\ell} \cap E) = (\partial B_\ell) * b_\ell$, and $f(\partial D_{2\ell} - \partial D_{2\ell} \cap E) = (\partial B_\ell) * c_\ell$, $\ell = 1,\dots,t$. Thus $r = f^{-1} h f$ induces a homeomorphism of V which in turn induces a homeomorphism of the union of Moore spaces in (7.51).

Any two disjoint framed imbeddings $\mathfrak{M}_t^k \to N$, $F|\mathfrak{M}_t^k(1)$ and $F|\mathfrak{M}_t^k(2)$, say, can be joined along $F(\bar{\mathfrak{M}}_t^k(i)) \subseteq F(\mathfrak{M}_t^k(i))$, where $\bar{\mathfrak{M}}_t^k(i)$ is the image $\bar{E}$ of E under the identifications in (7.51), $i = 1,2$. This is because the $F(\bar{\mathfrak{M}}_t^k(i))$ are in the "base-disc" (3.20). The resultant union $F(\mathfrak{M}_t^k(1)) \cup F(\mathfrak{M}_t^k(2))$ is the image of V and can be assumed to respect distinguished and complementary framings. Taking this union has no effect on (4C.14) or on ξ. The homeomorphism r constructed above thus induces a homeomorphism of the regular neighborhood of $F(\mathcal{C}_{tI_p}^k)$. Since the union of Figure (7.50) describes addition in $\pi_k(N;\mathbb{Z}_t)$, it is easily seen that r converts $\{F|\mathfrak{M}_t^k(1)\}$ to $-\{F|\mathfrak{M}_t^k(2)\}$ and $\{F|\mathfrak{M}_t^k(2)\}$ to $\{F|\mathfrak{M}_t^k(2)\} - \{F|\mathfrak{M}_t^k(1)\}$ where $\{\ .\ \}$ denotes homotopy class. Hence in $K_{k-1}(N)$ it converts x_1 to $-x_2$ and x_2 to $x_2 - x_1$. This together with (7.48)(a)-(b) is sufficient to give (7.48)(c) and completes the proof of (7.46). (For more details see [W2,§6].)

7.53 Proposition. Let $(g,b)\colon(N^{2k},\nu_N) \to (Y,\xi)$ be a highly connected local surgery problem and let $\underline{\underline{F}}_i = \begin{array}{ccc} \mathcal{C}^k_{\mu_i} & \xrightarrow{F_i} & N \\ \downarrow & & \downarrow g \\ \bar{\mathcal{C}}^{k+1}_{\mu_i} & \xrightarrow{\bar{F}_i} & Y \end{array}$ satisfy (7.5),

$i = 1,2$, where $\mu_1 \in M_n(\mathbb{Z}\pi,\mathbb{Q}\pi)^\times$ and $\mu_2 \in M_m(\mathbb{Z}\pi,\mathbb{Q}\pi)^\times$. Then

$$\{\theta(g,b;\underline{\underline{F}}_1,\mu_1)\} = \{\theta(g,b;\underline{\underline{F}}_2,\mu_2)\}$$

where $\{\theta\}$ denotes the equivalence class of the formation θ in $L_1^{(-1)^k}(\mathbb{Q}\pi/\mathbb{Z}\pi)$.

Proof. There exist $t \in \mathbb{Z}$, $\nu_1 \in M_n(\mathbb{Z}\pi,\mathbb{Q}\pi)^\times$, and $\nu_2 \in M_m(\mathbb{Z}\pi,\mathbb{Q}\pi)^\times$ such that $\mu_1\nu_1 = tI_n$ and $\mu_2\nu_2 = tI_m$. By (7.31)(a) we may assume $\mu_1 = tI_n$ and $\mu_2 = tI_m$. Further altering $\underline{\underline{F}}_1$ (resp. $\underline{\underline{F}}_2$) by adding in m (resp. n) disjointly imbedded discs as in (7.28), then stabilizing by

$$\begin{pmatrix} 0_n & 0 \\ 0 & tI_m \end{pmatrix} \quad \text{(resp.} \quad \begin{pmatrix} 0_m & 0 \\ 0 & tI_n \end{pmatrix}\text{)},$$

allows us to assume $\mu_1 = tI_p = \mu_2$, $p = m + n$. Further, since this stabilization adds m (resp. n) trivially imbedded copies of $\mathfrak{M}_t^k$ to $F_1(\mathcal{C}_{\mu_1}^k)$ (resp. $F_2(\mathcal{C}_{\mu_2}^k)$) and since $\pi_k(g) \cong K_{k-1}(N)$, we may assume by (7.46) that

7.54 $$\{\underline{\underline{F}}_1|(b\mathfrak{M}_t^{-k+1}(i),b\mathfrak{M}_t^k(i))\} = \{\underline{\underline{F}}_2|(b\mathfrak{M}_t^{-k+1}(i),b\mathfrak{M}_t^k(i))\}$$

where $\mathcal{C}_{\mu_1}^{-k+1} = \bigcup_{i=1}^{p} \mathfrak{M}_t^{-k+1}(i) = \mathcal{C}_{\mu_2}^{-k+1}$ is the disjoint union of relative Moore spaces, and $\{\ .\ \}$ denotes homotopy class (in $\pi_k(g)$).

Consider the commutative diagram of (3.15) where $s \in \mathbb{Z}$,

7.55 $$\begin{array}{ccccc} \pi_{k+1}(g)\otimes\mathbb{Z}_t & \xrightarrow{p_t} & \pi_{k+1}(g;\mathbb{Z}_t) & \xrightarrow{q_t} & \pi_k(g) \\ \downarrow m_s & & \downarrow r_s & & \downarrow = \\ \pi_{k+1}(g)\otimes\mathbb{Z}_{st} & \xrightarrow{p_{st}} & \pi_{k+1}(g;\mathbb{Z}_{st}) & \xrightarrow{q_{st}} & \pi_k(g) \end{array}$$

By the Serre-Hurewicz Theorem mod the class of finite abelian groups, $\pi_{k+1}(g)$ is finite, so for s large enough, $m_s = 0$. Let $\underline{\underline{F}}_{1j}$ and $\underline{\underline{F}}_{2j}$ denote the Moore space components of $\underline{\underline{F}}_1$ and $\underline{\underline{F}}_2$, $j = 1,\ldots,p$. Then (7.54) implies $q_t\{\underline{\underline{F}}_{1j}\} = q_t\{\underline{\underline{F}}_{2j}\}$ and with s chosen so that $m_s = 0$, (7.55) yields

7.56 $$r_s\{\underline{\underline{F}}_{1j}\} = r_s\{\underline{\underline{F}}_{2j}\}, \quad j = 1,\ldots,p.$$

By (4C.16), this application of r_s can be accomplished by stabilization of $\underline{\underline{F}}_1$ and $\underline{\underline{F}}_2$ by sI_p. Hence by (7.31)(a) we may assume $\underline{\underline{F}}_1 \simeq \underline{\underline{F}}_2$. By (7.24)(a), the proof of (7.53) is complete.

7.57 Definition. $\theta(g,b) \in L_1^{(-1)^k}(\mathbb{Q}\pi/\mathbb{Z}\pi)$ denotes the equivalence class of formations constructed in (7.53).

7.58. We next examine the effect of local surgery on $\theta(g,b;\underline{\underline{F}},\mu)$ in (7.15). Let the data of (7.4) and (7.5) be given and let $(G;B): (W,N',N;\nu_W) \to (Y\times I, Y\times\{1\}, Y\times\{0\}, \xi\times I)$ be the local bordism (4B.46) corresponding to surgery on $F(\mathcal{C}_\mu^k)$. By (4B.46)(ii), $(W,N') \approx (N\times I\cup\bar{U}, ((N-\mathring{U})\times\{1\})\cup V)$ where $\bar{U} = n\bar{\mathcal{C}}_\mu^{\prime,k+1}\times D^{k-1}$, $U = nF(\mathcal{C}_\mu^k)\times D^{k-1} \subseteq N$, $V = \partial\bar{U} - \mathring{U}$, and the first (resp. second) union is taken along $U\times\{1\}$ (resp. $\partial U\times\{1\}$) in $N\times\{1\}$ and $n\mathcal{C}_\mu^{\prime k}\times D^{k-1}$ in $\bar{U}$ (resp. ∂V in V). Thus,

7.59 $$(N_0', \partial V) = (N' - \mathring{V}, \partial V) = (N - \mathring{U}, \partial U) = (N_0, \partial U).$$

Let

7.60 $$I: \mathcal{C}_{\bar{\mu}}^k \longrightarrow \partial U \longrightarrow V$$

denote the composition of inclusions, where $\mathcal{C}_{\bar{\mu}}^k \hookrightarrow \partial U$ is the dual conglomerate associated to $\mathcal{C}_\mu^k$ in (4B.31). Using (4B.25) and the composition $u\bar{I}_*$ in (4B.69), it is easily shown that $\tilde{I}_*: H_*(\tilde{\mathcal{C}}_{\bar{\mu}}^k) \overset{\cong}{\to} H_*(\tilde{V})$ (with the obvious adaptation of notation from (4B.69)). Further, $\tilde{I}_\#: \pi_1\tilde{\mathcal{C}}_{\bar{\mu}}^k \overset{\cong}{\to} \pi_1\tilde{V}$ because one may show in (4B.31) that $\pi_1\tilde{\mathcal{C}}_{\bar{\mu}}^{k+1} \overset{\cong}{\to} \pi_1\tilde{U}$, while $\pi_1\tilde{U} \overset{\cong}{\leftarrow} \pi_1(\partial\tilde{U}) \equiv \pi_1(\partial\tilde{V}) \overset{\cong}{\to} \pi_1\tilde{V}$. Thus

7.61 $$I: \mathcal{C}_{\bar{\mu}}^k \xrightarrow{\simeq} V.$$

Now push $I(\mathcal{C}_{\bar{\mu}}^k)$ into $\mathring{V}$ along a collaring of ∂V in V and let $F': \mathcal{C}_{\bar{\mu}}^k \to N'$

denote the resultant framed imbedding. There is a corresponding square $\underline{\underline{F}}'$ (as in (7.4) satisfying the conditions of (7.5). Since $\pi_1\mathcal{C}^k_{\bar{\mu}}$ is a free group, (7.61) implies V is a regular neighborhood of $F'\mathcal{C}^k_{\bar{\mu}}$ in N'. (This could be proved directly from the constructions in (4B.31) and (4B.46)). Finally, if we push $(\bar{\mathcal{C}}'^{k+1}_{\mu}, \mathcal{C}^k_{\mu}) \subseteq (\bar{U},U)$ out to $(V,\partial V)$ by the first vector of its complementary framing, then it becomes the dual conglomerate pair (4B.31) to $F'\mathcal{C}^k_{\bar{\mu}} \hookrightarrow V$. (This also follows from the construction in (4B.31)).

Recall from (2.36)(b) the definition of $\omega\theta$, where θ is a formation, and keep the above notation for the following proposition.

7.62 Proposition. (a) Let $(g',b') = (G,B)|N'$. Then $\theta(g',b';\underline{\underline{F}}',\bar{\mu}) = \omega\theta(g,b;\underline{\underline{F}},\mu)$.

(b) If $(h,c)\colon (M,\nu_M) \to (Y,\xi)$ is any highly connected local surgery problem, locally bordant to (g,b), then there is $\mu \in M_n(\mathbb{Z}\pi,\mathbb{Q}\pi)^{\times}$ and a square $\underline{\underline{F}}$ (resp. $\underline{\underline{F}}'$) satisfying (7.5) with respect to (g,b) and μ (resp. (h,c) and $\bar{\mu}$) such that $\theta(h,c;\underline{\underline{F}}',\bar{\mu}) = \omega\theta(g,b;\underline{\underline{F}},\mu)$.

Proof. Let $(K_{k-1}(\partial V),\varphi,\psi)$ be the hyperbolic $(-1)^k$-form and let $b_1,\ldots,b_n$ (resp. $\bar{b}_1,\ldots,\bar{b}_n$) be the bockstein components of $\mathcal{C}^k_{\mu} \subseteq \partial U = \partial V$ (resp. $\mathcal{C}^k_{\bar{\mu}} \subseteq \partial U = \partial V$). We have the linking pairing (restriction of φ)

7.63 $$i'K_k(V,\partial V) \times s'K_{k-1}(V) \longrightarrow \mathbb{Q}\pi/\mathbb{Z}\pi$$

where $i'\colon K_k(V,\partial V) \to K_{k-1}(\partial V)$ is the homology boundary, $s'\colon K_{k-1}(V) \to K_{k-1}(\partial V)$ is the natural section (4B.45), and

7.64 $$K_{k-1}(\partial V) \cong i'K_k(V,\partial V) + s'K_{k-1}(V)\ .$$

After identifying $K_k(V,\partial V)$ with $K_{k-1}(V)^{\wedge}$ by duality, (7.63) is required to be the natural form (2.6)(c), since φ is hyperbolic (2.11)(b). Using the steps in the verification of (4B.44), this is equivalent to the requirement that (mod $\mathbb{Z}\pi$) $\varphi([b_i],[\bar{b}_j]) \equiv (\bar{\mu}^{-1})_{ij}$ = the (i,j)-entry in

$\bar{\mu}^{-1}$, $i,j = 1,\ldots,n$, since $V \approx n\mathcal{C}^k_{\bar{\mu}} \times D^{k-1}$. However, $\varphi([b_i],[\bar{b}_j]) = (-1)^k\, \overline{\varphi([\bar{b}_j],[b_i])} \equiv (-1)^k\, \overline{(\mu^{-1})_{ji}}$ (mod $\mathbb{Z}\pi$) where the last equality again uses (4B.44), this time viewing ∂V as ∂U. Since $\overline{(\mu^{-1})_{ji}} = (\bar{\mu}^{-1})_{ij}$,

$$\varphi([b_i],[\bar{b}_j]) = (-1)^k (\bar{\mu}^{-1})_{ij}. \tag{7.65}$$

To remove the factor $(-1)^k$ from (7.65), we reverse the orientation of $\mathcal{C}^k_\mu$ if k is odd. (This may be done using ρ in (7.49)). Let $\Delta' = (\gamma',\alpha')$ where $\Delta'\colon K_k(N_0',\partial V) \to K_{k-1}(\partial V)$ is the boundary map and (using (7.64)) $\gamma'\colon K_k(N_0',\partial V) \to K_k(V,\partial V)$ and $\alpha'\colon K_k(N_0',\partial V) \to K_{k-1}(V)$. By (4B.43), (7.61) and the above discussion there are isomorphisms

$$K_k(V,\partial V) \xleftarrow{\cong} H_k(\tilde{\mathcal{C}}'^{k+1}_\mu, \tilde{\mathcal{C}}^k_\mu) \xrightarrow{\cong} H_{k-1}(\tilde{\mathcal{C}}^k_\mu) \xrightarrow{\cong} K_{k-1}(U)$$

and

$$K_k(U,\partial U) \xleftarrow{\cong} H_k(\tilde{\mathcal{C}}^{k+1}_{\bar{\mu}}, \tilde{\mathcal{C}}^k_{\bar{\mu}}) \xrightarrow{\cong} H_{k-1}(\tilde{\mathcal{C}}^k_{\bar{\mu}}) \xrightarrow{\cong} K_{k-1}(V) .$$

Taking these as identifications and using (7.59)

$$\Delta' = (\gamma',\alpha') = (\epsilon\alpha,\gamma).$$

Hence it remains to verify that $\overline{\xi(x)} = \epsilon\xi'(x)$ for all $x \in K_k(N_0,\partial U) \equiv K_k(N_0',\partial V)$, where ξ (resp. ξ') is defined in (7.9) for (g,b) and $\underline{\underline{F}}$ (resp. (g',b') and $\underline{\underline{F}}'$) and $\epsilon = (-1)^k$. By (7.59) the framed immersion (7.10) $h\colon (\mathfrak{M}(x), b\mathfrak{M}(x)) \to (N_0,\partial U)$ can be used both for ξ and ξ', where (7.10)(b) now reads $[h(b\mathfrak{M}(x))] = [\gamma_x] + \epsilon[\alpha_x]$ (in the notation of (7.9)). Recalling that $\beta_x = \sum_{j,i} \mathfrak{M}(j)\nu_{ji}\alpha_i$ satisfies $\partial\beta_x = t\alpha_x$ (because $\mu\nu = tI_n$), it follows that $\delta_x = \sum_{j,i} {}_*\mathfrak{M}(j)\bar{\nu}_{ji}\gamma_i$ satisfies $\partial\delta_x = t\gamma_x$ (because $\bar{\mu}\bar{\nu} = tI_n$) where $\mathcal{C}^k_{\bar{\mu}} = \bigcup_j {}_*\mathfrak{M}(j)$. Hence using the argument in [W1,p.250]

$$\epsilon\, \overline{\gamma_x \pitchfork (\epsilon\beta_x)} = (\epsilon\alpha_x) \pitchfork \delta_x \tag{7.66}$$

where $\epsilon = (-1)^k$. Since $h\mathfrak{M}(x) \pitchfork h\mathfrak{M}(x) \in \mathbb{Z}\pi/J_{(-1)^{k+1}}$ where $J_{(-1)^{k+1}} = \{a - (-1)^k\bar{a} \mid a \in \mathbb{Z}\pi\}$,

7.67 $$\epsilon\,\overline{h\mathfrak{M}(x) \pitchfork h\mathfrak{M}(x)} = h\mathfrak{M}(x) \pitchfork h\mathfrak{M}(x).$$

Hence by (7.11), (7.66) and (7.67), $\xi'(x) = \epsilon\,\overline{\xi(x)}$ as required.

(b). Let $(G;B): (W,M,N;\nu_W) \to (Y\times I, Y\times\{1\}, Y\times\{0\};\xi\times I)$ be the local bordism. By (1.20)(b) (or rather its proof) we may assume that $K_i(W) = 0 = K_i(W,\partial W)$, $i \neq k$, without altering $G|\partial W = h \cup g$. A homology argument shows that $K_i(W,N) = 0$, $i \neq k$, so by a standard handle elimination argument (like that of [W2,p.61]) we may assume that $W = N \times I \cup$ (k-handles) $\cup$ ((k+1)-handles). By (4B.54) there is $\mu \in M_n(\mathbb{Z}\pi, \mathbb{Q}\pi)^{\times}$ such that W is the bordism corresponding to surgery on a framed imbedding $F: \mathcal{C}_\mu^k \to N$; F extends to $\underline{\underline{F}}$ (7.4) and is easily shown to satisfy (7.5)(a) (since $K_{k-1}(W) = 0$). By part (a) above, the proof of (7.62) is complete.

7.68. We can now prove the first statement in (7.3). By Prop. (7.53), $\theta(g,b) \in L_1^{(-1)^k}(\mathbb{Q}\pi/\mathbb{Z}\pi)$ depends only on (g,b). If (g,b) is locally bordant to a homotopy equivalence (g',b') then by (7.53) we may assume $\theta(g',b') = [\underline{\underline{0}}]$, the class of the zero formation. By (7.62) $\theta(g,b) = 0$. Conversely, if $\theta(g,b) = 0$, then some sequence of operations (2.36)(a)-(c) and their inverses converts the formation $\theta(g,b;\underline{\underline{F}},\mu)$ to $\underline{\underline{0}}$. By (7.24)(b), (7.31)(b)-(c), and (7.62)(b), all such operations are realizable by alterations in $\underline{\underline{F}}$ and μ (for fixed (g,b)) together with local bordisms of (g,b). Hence (g,b) is locally bordant to (g',b') where $\theta(g',b';\underline{\underline{F}}',\mu') = \underline{\underline{0}}$, for some $\underline{\underline{F}}'$ and μ'. This means g' is a homotopy equivalence. To complete the proof of (7.3) it remains to realize each element of $L_1^{\epsilon}(\mathbb{Q}\pi/\mathbb{Z}\pi)$ as a local surgery obstruction. This is done next.

7.69. Let $\theta \in L_1^{(-1)^k}(\mathbb{Q}\pi/\mathbb{Z}\pi)$ be represented by the formation (K,H,Δ,ξ) where $H \cong \operatorname{cok}\mu$ and $\mu \in M_n(\mathbb{Z}\pi,\mathbb{Q}\pi)^{\times}$. Let Q^{2k-2} be a closed manifold with $\pi_1 Q^{2k-2} = \pi$ and let $M^{2k-1} = Q^{2k-2} \times I$. Let $\bar{G}: \bar{\mathcal{C}}_\mu^k \to Q^{2k-2}$ be a framed imbedding (4B.58) and, using [HW,p.227], let $H: \mathcal{C}_\mu^k \times I \to M^{2k-1}$

be a framed immersion realizing some given list of type (a) circle self-intersections (see (5.10)(i) and (5.12)) for $H|\dot{\mathcal{C}}_\mu^{k-1} \times (I-\partial I)$, where $H|\mathcal{C}_\mu^{k-1} \times \{0\} = (\bar{G}|\mathcal{C}_\mu^{k-1}) \times \{\frac{1}{4}\}: \mathcal{C}_\mu^{k-1} \to M^{2k-1} \times \{\frac{1}{4}\}$ and $H|\mathcal{C}_\mu^{k-1} \times \{1\}: \mathcal{C}_\mu^{k-1} \to M^{2k-1} \times \{\frac{3}{4}\}$ is a framed imbedding. Let $g': (W', M', M, \partial M\times I) \to (M\times I, M\times\{1\}, M\times\{0\}, \partial M\times I)$ denote the bordism of the identity map $Id_M: M \to M$ corresponding to surgery on $H(\mathcal{C}_\mu^{k-1} \times \{1\}) \subseteq M$. (We omit here and later the obvious bundle maps from the notation. The reason for doing surgery on $H|\mathcal{C}_\mu^{k-1} \times \{1\}$ instead of on $H|\mathcal{C}_\mu^{k-1} \times \{0\} = \bar{G}|\mathcal{C}_\mu^{k-1}$ will become apparent below.)

By (4B.60),

$$7.70 \qquad K_i(M') = \begin{cases} H^{\wedge} + H, & i = k-1 \\ 0 & i \neq k-1 \end{cases} .$$

By (4B.46)(ii), $W' = M \times I \cup n\bar{\mathcal{C}}_\mu^{\prime k} \times D^{k-1}$; pushing $\bar{\mathcal{C}}_\mu^{\prime k}$ out to M' and regarding $H(\mathcal{C}_\mu^{k-1} \times I) \cup \bar{G}(\bar{\mathcal{C}}_\mu^{k})$ as a subspace of M' in the obvious way, it follows that $\bar{\mathcal{C}}_\mu^{\prime k} \cup H(\mathcal{C}_\mu^{k-1} \times I) \cup \bar{G}(\bar{\mathcal{C}}_\mu^{k})$ is the image of a framed immersion

$$F': \mathcal{C}_\mu^k \longrightarrow M'^{2k-1}$$

having the circle self-intersections of H. By (4B.46)(iii),

$$7.71 \qquad \operatorname{im}\{F'_*: H_{k-1}(\tilde{\mathcal{C}}_\mu^k) \to K_{k-1}(M')\} = H .$$

In the proof of (4B.60) a framed imbedding $I: \mathcal{C}_{\bar{\mu}}^k \to M'^{2k-1}$ was constructed (essentially the dual conglomerate (4B.31) corresponding to F') and was shown (4B.69) to satisfy

$$7.72 \qquad \operatorname{im}\{I_*: H_{k-1}(\tilde{\mathcal{C}}_{\bar{\mu}}^k) \to K_{k-1}(M')\} = H^{\wedge}.$$

It is easily verified that $I(\mathcal{C}_{\bar{\mu}}^k) \cap F'(\mathcal{C}_\mu^k)$ consists of line segments (of double points) connecting $F'(b\mathcal{C}_\mu^k)$ to $I(b\mathcal{C}_{\bar{\mu}}^k)$.

Let $E \overset{\mu}{\rightarrowtail} E' \overset{j}{\twoheadrightarrow} H$ be a free resolution of H and let $\bar{E}' \overset{\bar{\mu}}{\rightarrowtail} \bar{E} \overset{\tilde{j}}{\twoheadrightarrow} H^{\wedge}$ be its dual (2.15). Let D be the pull back of $\tilde{j} + j$ and Δ in

7.73
$$\begin{array}{ccc} \bar{E}' + E & \xrightarrow{=} & \bar{E}' + E \\ \downarrow \eta & & \downarrow \bar{\mu}+\mu \\ D & \xrightarrow{\rho} & \bar{E} + E' \\ \downarrow & & \downarrow \tilde{j}+j \\ K & \xrightarrow{\Delta} & H^{\wedge} + H \end{array} .$$

By Schanuel's Lemma D is stably free and the usual argument allows us to assume it is $\mathbb{Z}\pi$-free. Thus $(\bar{\mu} + \mu) = \rho\eta$, and if $I \cup F'\colon \mathcal{C}^k_{\bar{\mu}} \cup \mathcal{C}^k_{\mu} = \mathcal{C}^k_{\bar{\mu}+\mu} \to M'$ were an imbedding, (4C.19) would convert its image to $\mathcal{C}^k_{\rho}\mathcal{C}^k_{\eta}$ in M'. Letting $J\colon \mathcal{C}^k_{\eta} \to M$ denote the resultant inclusion, (7.73) would imply

7.74 $\quad K \cong H_{k-1}(\mathcal{C}^k_{\eta})$ and $\{\Delta\colon K \to H^{\wedge} + H\} = \{J_*\colon H_{k-1}(\mathcal{C}^k_{\eta}) \longrightarrow K_{k-1}(M')\}$.

However, the conversion in (4C.19) is accomplished by the operations of sliding up from the bockstein (4C.13), adding or subtracting 2-discs (4C.12) (these two cover Steps 1 and 2 in the proof of (4C.19)), adding copies of $S^{k-1} \times S^1$ along the bockstein (4C.38)(a)-(b), and adding in tubes (4C.38)(c)-(d), while the double point set of $I \cup F'$ is one-dimensional. Thus, since $k \geq 4$, we may carry out the required conversion to get a framed immersion $J\colon \mathcal{C}^k_{\eta} \to M'$ satisfying (7.74). Further, the double point set of J consists of line segments whose endpoint lie in $J(b\mathcal{C}^k_{\eta}) \cap J(\mathring{\mathcal{C}}^k_{\eta})$, together with circle intersections arising from F'. This is because $k \geq 4$ so we can assume that the only above-mentioned operation on $\mathcal{C}^k_{\bar{\mu}+\mu}$ which moves the intersection set is that of "sliding up from the bockstein." Since $\operatorname{im}(J_*) = \operatorname{im}(\Delta)$ is totally isotropic in $K_{k-1}(M') = H^{\wedge} + H$, we may assume by (4B.36) that $J(b\mathcal{C}^k_{\eta}) \cap J(\mathring{\mathcal{C}}^k_{\eta}) = \emptyset$. The next lemma describes the remaining double points.

7.75 <u>Lemma</u>. The self-intersection set of $J|\mathring{\mathcal{C}}^k_{\eta}$ consists of type (a) circles (5.10)(i).

This lemma will be proved below. It implies that by choosing an appropriate collection of circle self-intersections of H (thus of F') we may assume that the intersections created by application of (4B.36) cancel those already present in J. Thus we assume J is a framed imbedding satisfying (7.74).

Since $\mathrm{im}(J_*)$ is a subkernel of $K_{k-1}(M')$, a homology argument shows that surgery on $J(\mathcal{C}^k_\eta)$ yields a homotopy equivalence. (Compare (6.14) and the proof that surgery on a subkernel in the sense of [W2,§5] yields a homotopy equivalence; we cannot use (6.4).) More precisely let $g'': (W'',M'',M',\partial M\times I) \to (M\times I, M\times\{1\}, M\times\{0\}, \partial M\times I)$ be the bordism corresponding to surgery on $J(\mathcal{C}^k_\eta)$, where $W'' = M' \times I \cup n\bar{\mathcal{C}}'^{k+1}_\eta \times D^{k-2}$ (4B.46), $\bar{J}: \bar{\mathcal{C}}^{k+1}_\eta \to W''$ has image $\bar{\mathcal{C}}'^{k+1}_\eta$, and $g''|\partial M \times I = \mathrm{Id}_{\partial M} \times I$. Then $g''|M''$ is a homotopy equivalence and setting $W' \cup W'' = P^{2k}$, $M \times I = Z$, $g' \cup g'' = h$, $\xi = \nu_M \times I$, and $b:\nu_P \to \xi$ the natural bundle map,

$$(h;b):\ (P^{2k},\partial P;\nu_P) \longrightarrow (Z,\partial Z;\xi) \tag{7.76}$$

is a highly connected local surgery problem (with $h|\partial P$ a homotopy equivalence). A homology argument shows that

$$F:\ \mathcal{C}^k_\mu \xrightarrow{\ F'\ } M' \hookrightarrow P$$

satisfies (7.5)(b); (7.5)(a) is satisfied since F' has only type (a) circle self-intersections, so F is regularly homotopic to an imbedding (also denoted F) by the discussion in [W2,p.77]. If U denotes the regular neighborhood of $F\mathcal{C}^k_\mu$ in P, then by a slight modification of $\bar{J}$, $\bar{J}(\bar{\mathcal{C}}^{k+1}_\eta, \mathcal{C}^k_\eta) \subseteq (P_0, \partial U)$ where $P_0 = P - \mathring{U}$. A homology argument shows that $\bar{J}_*: H_k(\tilde{\bar{\mathcal{C}}}^{k+1}_\eta, \tilde{\mathcal{C}}^k_\eta) \xrightarrow{\cong} K_k(P_0,\partial U)$, and the commutative diagram

$$\begin{array}{ccc}
H_k(\tilde{\bar{\mathcal{C}}}^{k+1}_\eta, \tilde{\mathcal{C}}^k_\eta) & \xrightarrow[\cong]{\ \bar{J}_*\ } & K_k(P_0,\partial U) \\
\partial \downarrow \cong & & \downarrow \\
H_{k-1}(\tilde{\mathcal{C}}^k_\eta) & \longrightarrow & K_{k-1}(\partial U)
\end{array}$$

together with (7.71), (7.72), (7.70), (7.74) and the natural isomorphism $K_k(\partial U) \xrightarrow{\cong} H^\wedge + H$, show that

$$(K,H,\Delta) = (K_k(P_0,\partial U), K_{k-1}(\partial U), \{K_k(P_0,\partial U) \longrightarrow K_{k-1}(\partial U)\}).$$

Let $\xi': K_k(P_0,\partial U) \to \mathbb{Q}\pi/J_{(-1)^{k+1}}$ be the function constructed in (7.9) for (h,b). This may not coincide with the given ξ, so we need in general to modify it by modifying J. We begin with some algebraic observations. In case k is even (K,ζ,ξ) (see (2.34)) is skew-hermitian and it is easy to deduce from the relations in (2.5)(d) that ζ determines ξ. Hence $\xi = \xi'$ so the proof of (7.3) is complete in this case. If k is odd then by (2.5)(d), $\xi - \xi': K \to \mathbb{Z}\pi/J_{+1}$ $(= \ker\{r: \mathbb{Q}\pi/J_{+1} \to \mathbb{Q}\pi/\mathbb{Z}\pi\})$. There is an injection of abelian groups $\sum_{g\neq g^{-1}} (\mathbb{Q}/\mathbb{Z})_g + \sum_{g^2=1} (\mathbb{Q}/2\mathbb{Z})_g \to \mathbb{Q}\pi/J_{+1}$, where if $r_g \in \mathbb{Q}$ and $\{r_g\}$ denotes the class of r_g mod $\mathbb{Z}$ (resp. $2\mathbb{Z}$) when $g \neq g^{-1}$ (resp. $g^2 = 1$), the map sends $\{r_g\}$ to $r_g(g + g^{-1})$ if $g \neq g^{-1}$ (resp. to $r_g g$ if $g^2 = 1$). By (2.18), $\xi'(x)$ and $\xi(x)$ are in the image of this inclusion for each $x \in K$, so that we may assume that $\xi(x) - \xi'(x) \equiv \sum_{g^2=1} r_g g \pmod{J_{+1}}$ where $2r_g \in 2\mathbb{Z}$, so that $r_g \in \mathbb{Z}$.

7.77 Lemma. Let dim $M' = 2k$ where k is odd and let $\lambda_i \in \mathbb{Z}\pi$ be given, where $i = 1,\ldots,2n$ and each λ_i is of the form $\sum_{g^2=1} r_g g$, where $r_g = 0$ or 1. Then there is a homotopy $\Gamma: \mathcal{C}_\eta^k \times I \to M' \times I$ such that

a) $\Gamma|\mathcal{C}_\eta^k \times \{0\} = J$ and $\Gamma|\mathcal{C}_\eta^k \times \{1\}$ is a framed imbedding;

b) $\Gamma|b\mathcal{C}_\eta^k \times I$ is a regular homotopy; and

c) $\Gamma(b_i\mathcal{C}_\eta^k \times I) \pitchfork \Gamma(b_i\mathcal{C}_\eta^k \times I) = \lambda_i$, $i = 1,\ldots, 2n$.

Assuming this lemma, the proof of (7.3) is completed as follows.

Referring to the construction of g" above, instead of doing surgery on $J\mathcal{C}_\eta^k$, attach to M' a copy of $M' \times I$ (mapped to $M \times I$ by $(g'|M')\times I$). By the lemma and the discussion preceeding it, realize a collection of self-intersections so that

$$7.78 \qquad \Gamma(b\mathcal{C}_\eta^k \times I) \pitchfork \Gamma(b\mathcal{C}_\eta^k \times I) = \xi([b_i\mathcal{C}_\eta^k]) - \xi'([b_i\mathcal{C}_\eta^k]),$$

take $\Gamma(\mathcal{C}_\eta^k \times I) \subseteq M' \times I$ and do surgery on $\Gamma|\mathcal{C}_\eta^k \times \{1\}$. Once again we can construct (h,b) as in (7.76) and it follows from the definition (7.11) of ξ (for (h,b)) and (7.78) that $\theta(h,b)$ is represented by (K,H,Δ,ξ) as required. (Compare this procedure to that in the proof of [Sh 2,(10.1)(iii)].)

It remains to prove (7.75) and (7.77). We will be less precise here since some of these arguments have been seen before (notably in Chapter 5).

<u>Proof of</u> (7.75). All modifications of the intersection set in the proof of (4B.36) come from those in the proof of (4A.39). There are two steps in (4A.39): the first modifies $J|b\mathcal{C}_\eta^k$ (by a regular homotopy f realizing a given self-intersection number in $\mathbb{Z}\pi$ of $f|b\mathcal{C}_\eta^k \times I$ or else pushing $b\mathcal{C}_\eta^k$ along $\mathring{\mathcal{C}}_\eta^k$ as in (4A.29)) and the second applies the Whitney device. It is easily seen that after the first step, the double point set of J still consists of line segments. To use the Whitney device in the second step, let $\{p_1,p_2\} \subseteq J(b\mathcal{C}_\eta^k) \cap J(\mathring{\mathcal{C}}_\eta^k)$ be such that $g_{p_1} = g_{p_2}$ and $\epsilon_{p_1} = -\epsilon_{p_2}$ (see [W2, 5.2] for this notation). Suppose first that there are no line segments of double points connecting p_1 to p_2; then clearly the elimination of $\{p_1,p_2\}$ creates no circle intersections:

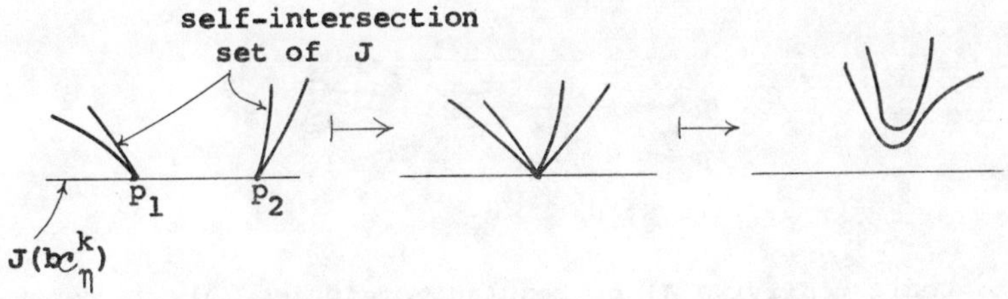

Suppose next that ℓ_2 is a line of double points of J connecting p_1 to p_2 and ℓ_1 is a path in $J(b\mathcal{C}_\eta^k)$ from p_2 to p_1 (not meeting any other double points). Then $\ell_2\ell_1$ is null-homotopic at p_1 since $g_{p_1} = g_{p_2}$. Thus we may span a 2-disc across $\ell_2\ell_1$ and push a neighborhood of ℓ_2 across it. In this way ℓ_2 is removed from the self-intersection set of J, while if ℓ_2' is another line of double points connecting p_1 to p_2, it is converted to a type (a) circle intersection of $J|\dot{\mathcal{C}}^k_\eta$, since $\ell_2'\ell_1$ is null-homotopic relative to $\{p_1\}$:

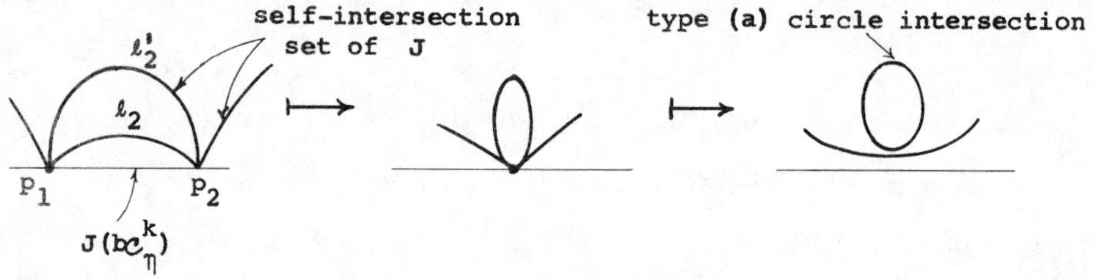

This is the inductive step in the removal (4B.46) of isolated points from $J(b\mathcal{C}^k_\eta) \cap J(\dot{\mathcal{C}}^k_\eta)$ so the proof of (7.75) is complete.

<u>Proof of</u> (7.77). It suffices to exhibit a homotopy satisfying conditions (a) and (b) for $i = 1$ and $\lambda_1 = g$ $(g^2 = 1)$ in (c). First let $\Gamma': \mathcal{C}^k_\eta \times I \to M' \times I$ be the regular homotopy of J constructed by pushing $b_1\mathcal{C}^k_\eta$ across itself with self-intersection number $g \in \mathbb{Z}\pi$ and let $J' = \Gamma'|\mathcal{C}^k_\eta \times \{1\}$. If m sheets abut to $b_1\mathcal{C}^k_\eta$ from $\dot{\mathcal{C}}^k_\eta$ the self-intersection set of J' consists of m^2 line segments whose endpoints may be divided into two disjoint subsets $P = \{p_1,\ldots,p_m\}$ and $Q = \{q_1,\ldots,q_m\}$ where each point of P is connected to each point of Q by some line segment:

7.79

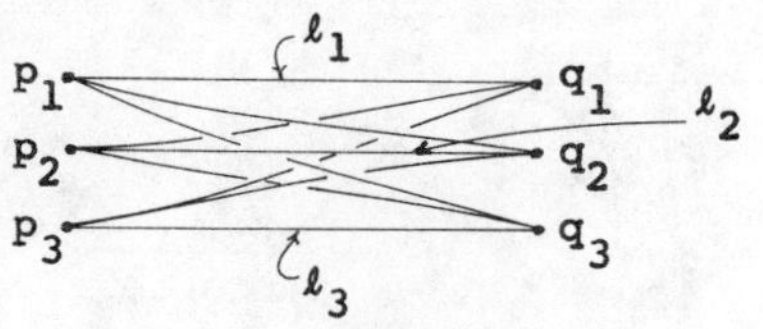

We will now begin modifying J' by regular homotopies, always denoting the result by J'. Choose paths ℓ_i of self-intersection of J' connecting p_i

to q_i, paths ℓ_i' in $J(b\mathcal{C}^k_\eta)$ connecting q_i to p_i, and let ℓ_i'' be ℓ_i' pushed into $J'(\overset{\bullet}{\mathcal{C}}{}^k_\eta)$, but keeping the endpoints fixed (in $J'(b\mathcal{C}^k_\eta)$), $i = 1,\dots,m$. Clearly we can span a 2-disc across $\ell_1''\ell_1'^{-1}$ and use the Whitney device to modify $J'|\overset{\bullet}{\mathcal{C}}{}^k_\eta$ removing $\{p_1,q_1\}$ from $J'(b\mathcal{C}^k_\eta) \cap J'(\overset{\bullet}{\mathcal{C}}{}^k_\eta)$. The line ℓ_1 is converted to a type (b) circle self-intersection of $J'|\overset{\bullet}{\mathcal{C}}{}^k_\eta$, spanning a loop homotopic to $\ell_1\ell_1'$ and hence contributing $g \in \mathbb{Z}_2\pi$ to the invariant of (5.10)(iii). (Compare [Sh 2, 8.2] for more details.) Figure (7.79) is converted to

7.80 <u>Figure</u>.

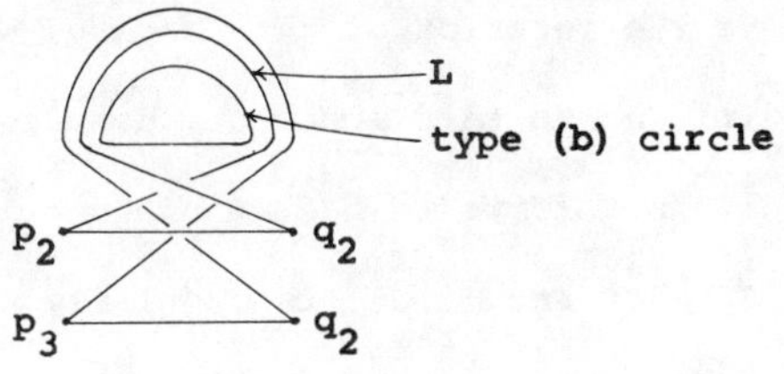

Applying the same procedure to eliminate $\{p_2,q_2\}$ yields another type (b) circle, while the segment L in Fig. (7.80) is converted to a <u>type (a)</u> circle because it (essentially) traverses ℓ_1' and ℓ_2' and so is homotopic to $g^2 = 1$. Hence, applying this construction to eliminate all points $P \cup Q = J'(b\mathcal{C}^k_\eta) \cap J'(\overset{\bullet}{\mathcal{C}}{}^k_\eta)$, we create m type (b) circles and also some type (a) circles in the double point set of $J'|\overset{\bullet}{\mathcal{C}}{}^k_\eta$. The type (a) circles may be eliminated as above by planting their inverses in J and then cancelling.

The above creation of type (b) intersections can be carried out "uniformly": there is a homeomorphism $h\colon C(m) \times S^{k-1} \xrightarrow{\approx} U$, a neighborhood of $b_1\mathcal{C}^k_\eta$, such that if $L_1,\dots,L_m \subseteq C(m)$ are the cone lines, then there are circles $\gamma_1,\dots,\gamma_m$, $\gamma_j \subseteq L_j \times S^{k-1}$, satisfying the following two conditions:

(a) the composition $\gamma_j \hookrightarrow L_j \times S^{k-1} \hookrightarrow C(m) \times S^{k-1} \xrightarrow{h} U \subseteq \mathcal{C}^k_\eta \xrightarrow{J'} M'$ is a double covering of its image; and

(b) let $c\colon C(m) \to R^2$ be the imbedding of (4A.5) and $i\colon S^{k-1} \to R^k$, the standard imbedding. Then there is a (k+2)-disc $B^{k+2} \subseteq R^{k+2}$ so that

$B^{k+2} \cap (c\times i)(L_j \times S^{k-1}) = D_j^k$, a k-disc containing γ_j and meeting $(c\times i)(0\times S^{k-1})$ in a $(k-1)$-disc $\subseteq \partial D_j^k$, where $0 \in C(m)$ is the cone point.

7.81 <u>Figure</u>.

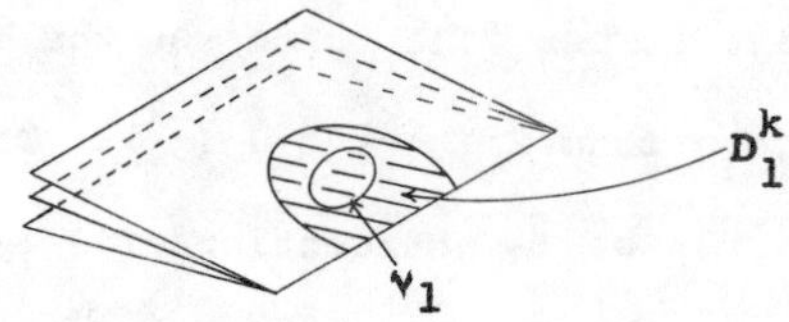

Now if we identify all the D_j^k to a single k-disc in B^{k+2} (by pushing them together in the vertical direction in the R^2-factor in $R^2 \times R^k = R^{k+2}$) and then remove the interior of this k-disc, what remains is still a copy of $\mathcal{C}_\eta^k$; J' restricted to this subspace, denoted J", has no type (b) circle self-intersections. (This is a form of "engulfing".)

Clearly $J' \simeq J''$ and $J'|b\mathcal{C}_\eta^k$ is regularly homotopic to $J|b\mathcal{C}_\eta^k$. J' and J" are not necessarily regularly homotopic, but this does not enter into the construction of ξ, since only $\Gamma|b\mathcal{C}_\eta^k$ is relevant. This completes the proof of (7.77).

<u>Remark</u>. In the construction of $\xi(x)$ (7.9), we used a representative of the form $h: (\mathfrak{M}(x), b\mathfrak{M}(x)) \to (P_0, \partial U)$ for x (7.10), where $\mathfrak{M}(x)$ is a Moore space. In (7.77) we implicitly claim that $\xi(x)$ can be computed from a representative for x of the form $\Gamma(b_i\mathcal{C}_\eta^k) \cup b_i\bar{\mathcal{C}}_\eta'^{k+1}$, where $\bar{\mathcal{C}}_\eta'^{k+1}$ is the relative Moore space added to M' for surgery on $\Gamma|\mathcal{C}_\eta^k \times \{1\}$ (4B.46). It is left to the reader to verify that this can be done.

Chapter Eight: The Simply-connected Case and π_1 Unrestricted

In this chapter we restate the main theorem (1.11) using the results of Chapters Six and Seven. We then compute its terms in the simply-connected case. Finally, we indicate how the main theorem may be proved without assuming finiteness of π_1.

Recall from (2.13) and (2.38) the definition of $L_i^{\epsilon}(\mathbb{Q}\pi/\mathbb{Z}\pi)$, $\epsilon = \pm 1$, $i = 0,1$.

8.1 Theorem. Let π be a finite group. For each integer n there is an exact sequence of abelian groups

$$\cdots \to L_n^h(\mathbb{Z}\pi) \to L_n^h(\mathbb{Q}\pi) \to L_{p(n)}^{(-1)^{[\frac{n}{2}]}}(\mathbb{Q}\pi/\mathbb{Z}\pi) \to L_{n-1}^h(\mathbb{Z}\pi) \to L_{n-1}^h(\mathbb{Q}\pi) \to \cdots$$

where $p(n) = \begin{cases} 0 \ , & n \text{ even} \\ 1 \ , & n \text{ odd} \end{cases}$ and $[\cdot]$ = integral part.

Proof. See (1.11), (6.1) and (7.1).

8.2 Remark. By an algebraic translation of the geometric results of this paper, one may derive a more general version of the exact sequence (see [P3]), in which $\mathbb{Z}\pi$ (resp. $\mathbb{Q}\pi$) is replaced by a ring A (resp. by a suitable ring of fractions B). It is in this context that the theorem has been used in [P5] to make calculations of $L_*(\mathbb{Z}\pi)$, for certain finite π.

It is difficult to apply (8.1) directly to such calculations, essentially because $\mathbb{Z}\pi$ is of infinite global dimension. There are, however, interesting cases where this difficulty is avoided. For example, if $\mathbb{Z}\pi$ is replaced by $\mathbb{Z}_{(p)}\pi$ in (8.1), so that the local term becomes $L_i^{\epsilon}(\mathbb{Q}\pi/\mathbb{Z}_{(p)}\pi)$ (a Grothendieck group of forms on p-torsion modules in $\mathfrak{D}_F^1$), the sequence remains exact. If $(p,|\pi|) = 1$, then $\mathbb{Z}_{(p)}\pi$ has global dimension one; or if π is the non-trivial split extension of an odd order cyclic group C by $\mathbb{Z}/2\mathbb{Z}$ ($C \rightarrowtail \pi \twoheadrightarrow \mathbb{Z}/2\mathbb{Z}$), then $\mathbb{Z}_{(2)}\pi$ is a product of hereditary (hence global dimension one) rings and a single $\mathbb{Z}_{(2)}[\mathbb{Z}/2\mathbb{Z}]$-factor. These examples occur in the work of T. Petrie [Pe]; here the local terms in (8.1) can be computed and, in turn, yield an essentially complete description of $L_*^h(\mathbb{Z}_{(p)}\pi)$. The main points in this computation are encountered in the simply-connected case ($\mathbb{Z}$ has homological dimension one). This is discussed next.

8.3. To carry out the description of (8.1) in case $\pi = 1$ and to make explicit the connection between the localization sequence and classical invariants for bilinear forms, we recall in detail some well-known theorems about Witt groups. Let R denote $\mathbb{Z}$ or a field (of any characteristic). Following [MH], $W(R)$ denotes the Grothendieck group of isometry classes of R-valued, symmetric, bilinear forms on free R-modules, modulo the subgroup generated by metabolic forms. (A form $\varphi\colon R^n \times R^n \to R$ is metabolic if $n = 2k$, and there is a direct summand $S \subseteq R^n$ of rank k such that $\varphi|S \times S \equiv 0$.) If $\frac{1}{2} \in R$, then each metabolic form is hyperbolic; thus, for example, $W(\mathbb{Q}) \cong L_{4k}(\mathbb{Q})$.

Given $r \in R$, let $\langle r \rangle$ denote the class in $W(R)$ of the unary (one-dimensional) form with matrix r. There is a map

$$\sigma_*\colon W(\mathbb{Q}) \longrightarrow W(\mathbb{Z})$$

defined by $\sigma_*\varphi = \sigma(\varphi)\langle 1 \rangle$, where $\sigma(\varphi) \in \mathbb{Z}$ is the signature of the form φ.

Let $p, r \in \mathbb{Z}$ be given and let $r = p^n r'$ where $(r', p) = 1$. For each prime p there is a map (where $\mathbb{F}_p$ is the field of p elements)

$$\partial_p\colon W(\mathbb{Q}) \longrightarrow W(\mathbb{F}_p)$$

defined on unary forms $\langle r \rangle$, $r \in \mathbb{Z}$, by

8.4 $$\partial_p\langle r\rangle = \begin{cases} \langle r' \bmod p\rangle , & n \text{ odd} \\ 0 , & n \text{ even.} \end{cases}$$

Given an arbitrary symmetric bilinear form over $\mathbb{Q}$, one may diagonalize and clear out denominators in its matrix representation, so that every form over $\mathbb{Q}$ is isometric to an orthogonal sum of forms $\langle r\rangle$ as above. One may also show that ∂_p sends metabolic forms to metabolic forms, so (8.4) defines ∂_p. Finally there is an obvious map $W(\mathbb{Z}) \to W(\mathbb{Q})$.

8.5 Proposition [MH, IV.2.1, IV.2.7, II.4.4, IV.1.5]. The sequence of abelian groups

$$W(\mathbb{Z}) \rightarrowtail W(\mathbb{Q}) \xrightarrow{\Sigma\partial_p} \sum_{p \text{ prime}} W(\mathbb{F}_p) \qquad (\sigma_*: W(\mathbb{Q}) \dashrightarrow W(\mathbb{Z}))$$

is short exact and split by σ_*. Further, there are isomorphisms

$$W(\mathbb{Z}) \xrightarrow{\cong} \mathbb{Z}$$

induced by the signature and

$$W(\mathbb{F}_p) \cong \begin{cases} \mathbb{Z}/2\mathbb{Z} + \mathbb{Z}/2\mathbb{Z} , & p \equiv 1(4) \\ \mathbb{Z}/4\mathbb{Z} , & p \equiv 3(4) \\ \mathbb{Z}/2\mathbb{Z} , & p = 2 \end{cases}$$

induced by rank and discriminant.

8.6 Proposition. $L_i(\mathbb{Q}) \cong \begin{cases} W(\mathbb{Z}) + \sum_{p \text{ prime}} W(\mathbb{F}_p), & i \equiv 0(4) \\ 0 , & \text{otherwise} \end{cases}$

Proof. If $i \equiv 0(4)$, $W(\mathbb{Q}) \equiv L_i(\mathbb{Q})$. To prove $L_{4k+2}(\mathbb{Q}) = 0$, use the classical fact that a skew-symmetric form over a field of characteristic $\neq 2$ is hyperbolic (admits a symplectic basis). If i is odd, use either the geometric argument of [Br, IV.2.14] and the remark following (1.11); or the algebraic argument of [P4].

8.7 Theorem.

$$L_i^{\epsilon}(\mathbb{Q}/\mathbb{Z}) \cong \begin{cases} \mathbb{Z}/8\mathbb{Z} + \sum\limits_{p \text{ prime}} W(\mathbb{F}_p), & i = 0,\ \epsilon = 1 \\ 0, & i = 0,\ \epsilon = -1 \\ 0, & i = 1,\ \epsilon = 1 \\ \mathbb{Z}/2\mathbb{Z}, & i = 1,\ \epsilon = -1. \end{cases}$$

Proof. To treat the cases where $\epsilon = 1$, consider the diagram (for any integer k),

8.8 Diagram.

$$\begin{array}{ccccccc} L_1^1(\mathbb{Q}/\mathbb{Z}) & \rightarrowtail & L_{4k}(\mathbb{Z}) & \longrightarrow & L_{4k}(\mathbb{Q}) & \xrightarrow{\delta} & L_0^1(\mathbb{Q}/\mathbb{Z}) \\ & & \downarrow a & & = \downarrow b & & \downarrow c \\ & & W(\mathbb{Z}) & \rightarrowtail & W(\mathbb{Q}) & \xrightarrow{\Sigma\partial_p} & \sum\limits_{p \text{ prime}} W(\mathbb{F}_p). \end{array}$$

Here the top line is exact by (8.1), (8.6) and the theorem of Kervaire-Milnor that $L_{\pm 1}(\mathbb{Z}) = 0$ [Br, II.1.1]; the bottom line is (8.5); a and b are defined by forgetting that forms are even; c is the induced map; and δ is defined geometrically by requiring that the diagram

$$\begin{array}{ccc} {}^1L_{4k}^{\mathbb{Q}}(K) & \xrightarrow{d} & {}^1L_{4k-1}^{\mathbb{Q}/\mathbb{Z}}(K) \\ \cong \downarrow \sigma_{\mathbb{Q}} & & \cong \downarrow (6.1) \\ L_{4k}(\mathbb{Q}) & \xrightarrow{\delta} & L_0^1(\mathbb{Q}/\mathbb{Z}) \end{array}$$

commute, where the top line is from Thm. (1.11) and $\pi_1(K) = \{1\}$.

To obtain an algebraic description of δ, it suffices (as in the discussion of ∂_p above) to evaluate $\delta\langle r\rangle$, $r \in \mathbb{Z}$. Thus, let the normal map $(f;b)\colon (M^{4k}, \partial M^{4k}; \nu_M) \to (D^{4k}, S^{4k-1}; \epsilon)$ be obtained by plumbing according to the (1×1)-matrix $[r]$ (r must be chosen even and $k \geq 2$--see [Br, V.2.1]). By definition, $(f;b)$ represents $(\sigma_{\mathbb{Q}})^{-1}\langle r\rangle$. Since $d(f;b) = (f|\partial M; b|\partial M)$,

8.9 $$\delta\langle r\rangle = [K_{2k-1}(\partial M), \varphi_r, \psi_r]$$

where φ_r and ψ_r are the linking and self-linking forms on $K_{2k-1}(\partial M)$. It is easily seen that $K_{2k-1}(\partial M) \cong \mathbb{Z}/r\mathbb{Z}$, with a generator g such that $\varphi_r(g,g) = \frac{1}{r} \in \mathbb{Q}/\mathbb{Z}$ and $\psi_r(g) = \frac{1}{r} \in \mathbb{Q}/2\mathbb{Z}$.

The signature induces an isomorphism $W(\mathbb{Z}) \xrightarrow{\cong} \mathbb{Z}$ and so by [Br, III.3.11, V.2.9], a is a monomorphism whose image is the ideal of integers divisible by 8. Thus Diagram (8.8) shows $L_1^1(\mathbb{Q}/\mathbb{Z}) = 0$ and yields the short exact sequence

$$8.10 \qquad \mathbb{Z}/8\mathbb{Z} \xrightarrow{i} L_0^1(\mathbb{Q}/\mathbb{Z}) \xrightarrow{j} \sum_{p \text{ prime}} W(\mathbb{F}_p)$$

where the generator of $\mathbb{Z}/8\mathbb{Z}$ is sent to $\delta\langle 1\rangle = \delta\langle 4\rangle = [\mathbb{Z}/4\mathbb{Z}, \varphi_4, \psi_4] \in L_0^1(\mathbb{Q}/\mathbb{Z})$ (see (8.9)).

To split (8.10), let (T,φ,ψ) be any symmetric linking form and (following [MS]) set

$$8.11 \qquad G(T,\varphi,\psi) = |T|^{-\frac{1}{2}} \sum_{t \in T} e^{\pi i \psi(t)} .$$

By [MS, 5.9] $G(T,\varphi,\psi)$ is an eighth root of unity which equals one if (T,φ,ψ) is a kernel [MS, p. 507, Lemma (iv)]. An easy computation shows that $G((T,\varphi,\psi) \perp (T',\varphi',\psi')) = G(T,\varphi,\psi)\cdot G(T',\varphi',\psi')$, so G induces a homomorphism

$$8.12 \qquad \Gamma\colon L_0^1(\mathbb{Q}/\mathbb{Z}) \longrightarrow \mathbb{Z}/8\mathbb{Z}.$$

Γ is right-inverse to i in (8.11) because $G((\mathbb{Z}/4\mathbb{Z}, \varphi_4, \psi_4)) = e^{\pi i/4}$, a primitive eighth root. This completes the discussion of the case $\epsilon = 1$ in (8.7).

From (8.1) and (8.6) follow the stated values of $L_i^{-1}(\mathbb{Q}/\mathbb{Z})$. To exhibit the role of ξ in the definition of "formation" (see (2.34)), we will sketch the construction of the non-trivial element of $L_1^{-1}(\mathbb{Q}/\mathbb{Z})$.

Let $(f;b)\colon (M^{4k+2}, \partial M; \nu_M) \to (Y,X;\xi)$ be a normal map of simply-connected spaces where $K_*(\partial M) = 0$, $K_i(M) = 0$ $(i \neq 2k+1)$, and $K_{2k+1}(M) = \mathbb{Z} + \mathbb{Z}$, with basis $\{e,f\}$. Let $\lambda(e,e) = 0 = \lambda(f,f)$, $\lambda(e,f) = 1$, and $\mu(e) = 1 = \mu(f)$, where λ and μ are the intersection and self-intersection forms of [W2, §5]. Thus $(f;b)$ represents the non-trivial element of $L_2(\mathbb{Z})$. Surgery on a framed imbedded sphere representing 2e yields a local surgery problem $(g;b): (N^{4k+2}, \partial N; \nu_N) \to (Y,X;\xi)$ where $K_{2k+1}(N) \cong \mathbb{Z}/2\mathbb{Z} \cong K_{2k}(N)$. The non-zero element $x \in K_{2k}(N)$ cannot be represented (in the sense of (3.17) and (4A.21)) by a framed embedding $h': \mathfrak{M}_2^{2k+1} \to N$ because surgery on h' would yield a homotopy equivalence. In fact, since $\mu(f) = 1$, it is easily seen from the construction of (g;b) that h' can be taken to have a single self-intersection. Thus, by (4A.35) we may double the order of the bockstein to find a framed imbedded representative $h: \mathfrak{M}_4^{2k+1} \to N$ for x. Referring to Diagram (7.7), we have $K_{2k}(U) = \mathbb{Z}/4\mathbb{Z}$, which forces $|K_{2k+1}(N,U)| = 4$. Suppose $K_{2k+1}(N,U) = \mathbb{Z}/4\mathbb{Z}$. Since $\Delta = (\gamma,\alpha)$ is injective in (7.7) and α must be multiplication by 2, γ is an isomorphism. We could thus do surgery on $h(\mathfrak{M}_4^{2k+1})$ to get a homotopy equivalence (see (7.58) and (7.62)). Hence $K_{2k+1}(N,U) \cong \mathbb{Z}/2\mathbb{Z} + \mathbb{Z}/2\mathbb{Z}$, and we may find generators u and v for it so that $\Delta: K_{2k+1}(N,U) \to K_{2k}(U)^\wedge + K_{2k}(U)$ is $i + i: \mathbb{Z}/2\mathbb{Z} + \mathbb{Z}/2\mathbb{Z} \to \mathbb{Z}/4\mathbb{Z} + \mathbb{Z}/4\mathbb{Z}$, the inclusion in each factor. It is left to the reader to construct geometric representatives for u and v, and following the construction of ξ in (7.9), show that $\xi(u) = \xi(v) = 1 \in \mathbb{Q}/2\mathbb{Z}$.

Finally, observe that if the formation just constructed is changed by setting $\xi(u) = 0$ (and $\xi(v) = 0$ or 1) then it represents zero in $L_1^{-1}(\mathbb{Q}/\mathbb{Z})$. In fact, this new formation may be destabilized to $\theta = (K,H,\Delta,\xi)$ where $K \cong \mathbb{Z}/2\mathbb{Z} \cong H$, $\Delta = (\gamma,\alpha): \mathbb{Z}/2\mathbb{Z} \xrightarrow{(id,0)} \mathbb{Z}/2\mathbb{Z} + \mathbb{Z}/2\mathbb{Z}$, and $\xi \neq 0$; $\omega\theta$ (see (2.36)(b)) may then be destabilized to the zero formation. This completes the proof of (8.7).

8.13. In each chapter of this paper but the first it is assumed that the fundamental group of a surgery problem is finite; and in Chapter Two we treat forms over $\mathbb{Z}\pi$-modules, where π is finite. Although this restrictive assumption is sufficient for the applications in [P3] and [P5] it has been pointed out by several people that it is unnecessary. (This was also not realized in [W1, §5], where it was conjectured that it was necessary to use the injective hull of $\mathbb{Z}\pi$ in place of $\mathbb{Q}\pi$. When π is finite $\mathbb{Q}\pi$ is the injective hull.) We will sketch here some of the arguments which remove the finiteness restriction; full details are available on request from the author.

None of the geometric arguments used finiteness of π_1. For example, all the geometric constructions of Chapter Four carry over verbatim to the general case. Hence we will mention here only the (algebraic) arguments of Chapter Two which must be changed.

No definitions need be changed. Clearly Props. (2.4)(a),(d) are false in general; (2.4)(c) remains valid if $M' \rightarrowtail M \twoheadrightarrow M''$ is a short exact sequence in $\mathfrak{D}_F^1$ and may be proved by resolving the sequence as in [CE,V.2] and applying $\mathrm{Hom}(—,B/A)$ to the resultant commutative ladder. Prop. (2.8) is proved as is [W2, Lemma 2.3(a)], except that we assume $H^{k+2}(C_*;B) = 0$ for any $\mathbb{Z}\pi$-module B (in place of $H^{k+1}(C_*;B) = 0$) and obtain a short (stably-) free resolution for $H_k(C_*)$ (in place of a splitting of $Z_k \twoheadrightarrow H_k(C_*)$). The assumption $H^{k+2}(C_*;B) = 0$ is justified by duality; the proof of (2.8) in Chapter Two does not use duality. In Example (2.10) the non-singularity of the linking pairing can be proved by an argument recognizing the groups and maps of [W2, p. 56] in terms of a "covering" of φ, in the sense of (2.19). Remark (2.12)(a) uses only (2.4)(c), as amended above; in (2.12)(b), $|T|$ is not generally definable in a useful way. (In later chapters $|\cdot|$ is applied several times to torsion modules T; this can be avoided by more careful duality arguments.) For (2.15), observe that if $M \in \mathfrak{D}_F^1$, then M is finitely generated (over $\mathbb{Z}\pi$). Thus, $\mathrm{Ext}_A^1(M,B) = 0$ because there is $t \in \mathbb{Z}$ such that $tM = 0$, while multiplication by t induces an isomorphism of B. In the argument of Theorem (2.22), we need only know in general that

$(J^{\perp}/J, \rho_1, \sigma_1)$ is non-singular (2.29); this follows from (2.31) where it is shown that $(J^{\perp}/J, \rho_1, \sigma_1)$ is isometric to $(V, \zeta, \xi) \perp (S, \varphi, \psi)$ (and non-singularity is not used).

Department of Mathematics
Columbia University

Index of Symbols and Terminology

$^1L_n^{\mathbb{Q}/\mathbb{Z}}(K)$ ... 1.5

$^2L_n^{\mathbb{Q}/\mathbb{Z}}(K)$ ... 1.6

$^1L^{\mathbb{Q}}_n(K)$ ... 1.8

$^iL_n^{\mathbb{Z}}(K)$, $^2L_n^{\mathbb{Q}}(K)$, ... 1.9

Local surgery problem and local bordism ... 1.12

Highly-connected local surgery problem ... 1.21

(A,B) means ($\mathbb{Z}\pi,\mathbb{Q}\pi$) in Chapter Two.

τM (torsion part of M) ... 2.2

fM (free part of M) ... 2.2

sesquilinear
nonsingular
ϵ-hermitian
ϵ-linking form } ... 2.5

natural form ... 2.6(c)

J_ϵ ... 2.5

ϵ-form ... 2.9

kernel, hyperbolic, $\mathscr{H}(U)$ ... 2.11

$\bar{M}, M^\wedge$ ... 2.2

covering ... 2.19

$F_0^\epsilon(B/A), \mathscr{F}_0^\epsilon(B/A), L_0^\epsilon(B/A)$ ... 2.13

$\mathfrak{D}_F^1$ (torsion modules which have a short free resolution) ... 2.7

ϵ-formation, $F_1^\epsilon(B/A)$ ... 2.34

$L_1^\epsilon(B/A)$ ... 2.38

$\chi_{(H,\beta,\sigma)}{}^\theta$
$\omega\theta$
$\sigma_E\theta$ } ... 2.36

$M_n(A,B)^\times$ ... 2.44

$E(\nu,\mu)$ ... 2.44

$\mathfrak{M}^k_{\{t_1,\ldots,t_n\}}$ (Generalized Moore space)
$b\mathfrak{M}^k_{\{t_1,\ldots,t_n\}}$ (bockstein)
$\mathfrak{M}^k_t$ (Moore space)
$\mathring{\mathfrak{M}}^k_t$ (interior) } 3.1, 4A.1

$\bar{\mathfrak{M}}^{k+1}_{\{t_1,\ldots,t_n\}}$ (relative generalized Moore space)
$b\bar{\mathfrak{M}}^{k+1}_{\{t_1,\ldots,t_n\}}$ (bockstein)
$\mathring{\mathfrak{M}}^{k+1}_{\{t_1,\ldots,t_n\}}$ (interior)
$\bar{\mathfrak{M}}^{k+1}_t$ (relative Moore space) } 3.7, 4A.1

$\pi_{k+1}(Y,X;\mathbb{Z}_t)$ ... 3.13

$\pi_{k+1}(\Phi;\mathbb{Z}_t)$ ... 3.18

$c\bar{\mathfrak{M}}^{k+1}_t, c\mathfrak{M}^k_t$ (collar) ... 4A.1

C(t) (cone on t points) ... 4A.3,4A.4

c: $C(t) \to \mathbb{R}^2$ ... 4A.5, 4A.6

$\nu_{C(t)}$ ... 4A.5, 4A.7

immersion of a Moore space
framed Moore space
regular homotopy
normal bundle, trivial bundle } 4A.15

ϵ (distinguished field),
complementary framing } ... 4A.16

$({}_*\bar{\mathfrak{M}}^{m+2}_t, {}_*\bar{\mathfrak{M}}^{m+1}_t)$ (dual Moore space pair) ... 4A.19

$\pitchfork$ ... 4A.27

$\pitchfork\!\!\pitchfork$ ... 4A.28

increasing the order of the bockstein ... 4A.35

$\tilde{\mathcal{C}}^k_\mu, b\tilde{\mathcal{C}}^k_\mu, \mathcal{C}^k_\mu, b\mathcal{C}^k_\mu$ (conglomerates) .. 4B.1

$\mathfrak{M}(i)$ (identified Moore space) ... 4B.13

$H_*(K), K \subseteq M$ ("local coefficients") ... 4B.17

$\tilde{\tilde{\mathcal{C}}}^{k+1}_\mu, \bar{\mathcal{C}}^{k+1}_\mu, b\tilde{\tilde{\mathcal{C}}}^{k+1}_\mu, b\bar{\mathcal{C}}^{k+1}_\mu$ (relative conglomerates)... 4B.21, 4B.22

Dual conglomerate ... 4B.31

Surgery on a conglomerate ... 4B.53

trivially imbedded conglomerate ... 4B.59

$\tilde{\mathcal{C}}^k_\mu \tilde{\mathcal{C}}^k_\nu, \mathcal{C}^k_\mu \mathcal{C}^k_\nu$ (composition of conglomerates) ... 4C.2

type(a), type(b) intersections ... 5.12

$\rightarrowtail$ ... monomorphism

$\twoheadrightarrow$ ... epimorphism

$\Sigma, M+N$ ... direct sum

[*] means a bibliographical reference to *, and (*) refers to the item (theorem, lemma, diagram, etc.) * in this paper.

Bibliography

[Ba] H. Bass, "Unitary algebraic K-theory" Lecture Notes in Mathematics, Springer Verlag No. 343 (1973), 57-265.

[Br] W. Browder, Surgery on Simply-Connected Manifolds (Bd. 65, Ergebnisse der Mathematik und ihre Grenzgebiete), Springer-Verlag, 1972.

[BRS] S. Buoncristiano, C.P. Rourke, and B.J. Sanderson, A Geometric Approach to Homology Theory, Cambridge University Press, 1976.

[CE] H. Cartan and S. Eilenberg, Homological Algebra, Princeton University Press, 1956.

[CM] G. Carlsson and R.J. Milgram, "Some exact sequences in the theory of bilinear and quadratic forms", preprint.

[C] F.X. Connolly, "Linking numbers and surgery", Topology 12 (1973) 389-412.

[CS] S. Cappell and J. Shaneson", The codimension-two placement problem and homology equivalent manifolds", Ann. of Math. 99 (1974), 277-348.

[Gr] D. Grayson, "Higher algebraic K-theory II (after D. Quillen)", Lecture Notes in Mathematics, Springer-Verlag No. 551 (1976) 217-240.

[H] A. Haefliger, "Lissage des Immersions I", Topology 6 (1967) 221-239

[Ha] J.-C. Hausmann, "On odd-dimensional surgery with finite fundamental group", to appear in Proc. Amer. Math. Soc.

[Hi] P. Hilton, Homotopy Theory and Duality, Gordon and Breach, 1965.

[HP] A. Haefliger and V. Poenaru, "La classification des immersions combinatoires", Publ. Math. I.H.E.S. 23 (1964) 75-91.

[HQ] A. Hatcher and F. Quinn, "Bordism invariants of intersections of manifolds", Trans. Amer. Math. Soc. 200 (1975) 327-344.

[HR] A. Heller and I. Reiner, "Grothendieck groups of orders in semi-simple algebras", Trans. Amer. Math. Soc. 112 (1964) 344-355.

[HW] A. Hatcher and J. Wagoner, "Pseudo-isotopies of compact manifolds", Asterisque 6 (1973) 1-238.

[K] M. Karoubi, "Localisation des formes quadratiques I, II", Ann. Scient. Ec. Norm. Sup. 7 (1974) 359-404; ibid 8 (1975) 99-155.

[Ma] S. Mac Lane, Homology, Springer-Verlag, 1963.

[MH] J. Milnor and D. Husemoller, Symmetric Bilinear Forms (Bd. 73, Ergebnisse der Mathematik und ihre Grenzgebiete), Springer-verlag, 1973.

[N] I. Namioka, "Maps of pairs in homotopy theory", Proc. Lond. Math. Soc. 12 (1962), 725-738.

[Nov] S.P. Novikov, "The algebraic construction and properties of Hermitian analogues of K-theory for rings with involution, from the point of view of Hamiltonian formalism. Some applications to differential topology and the theory of characteristic classes", I. Izv. Akad. Nauk SSR, ser math 34 (1970) 253-288. II. ibid 34 (1970) 475-500.

[P1] W. Pardon, "The exact sequence of a localization in L-theory", Ph.D. Thesis, Princeton University (1974).

[P2] W. Pardon, "Local surgery and some applications to the theory of quadratic forms", Bull. AMS 82 (1976) 131-133.

[P3] W. Pardon, "The exact sequence of a localization for Witt groups", Lecture Notes in Mathematics, Springer-Verlag No. 551 (1976) 336-379.

[P4] W. Pardon, "An invariant determining the Witt class of a unitary transformation over a semi-simple ring", Journal of Algebra 44 (1977) 396-410.

[P5] W. Pardon, "The exact sequence of a localization for Witt groups: applications", preprint.

[Pe] T. Petrie, Oral communication.

[R] D. Rim, "Modules over finite groups", Ann. of Math. 69 (1959) 700-712.

[Ra 1] A. Ranicki, "Foundations of algebraic L-theory I", Proc. Lond. Math. Soc. 27 (1973) 101-125.

[Ra 2] A. Ranicki, "The algebraic theory of surgery", preprint.

[Sh 1] R. Sharpe, "On the structure of the unitary Steinberg group", Ann. of Math. 96 (1972) 444-479.

[Sh 2] R. Sharpe, "Surgery on compact manifolds: the bounded even-dimensional case", Ann. of Math. 98 (1973) 187-209.

[Sm] J. Smith, "Complements of codimension-two submanifolds", Ph.D. Thesis, New York University (1976).

[Sp] E. Spanier, Algebraic Topology, Mc Graw-Hill, 1966.

[St] N. Stoltzfus, "Unravelling the integral knot concordance group", preprint.

[Sz] G. Szekeres, "Determination of a certain family of finite metabelian groups", Trans. Amer. Math. Soc. 66 (1949) 1-43.

[W1] C.T.C. Wall, "Surgery of non-simply-connected manifolds", Ann. of Math. 84 (1966) 217-276.

[W2] C.T.C. Wall, Surgery on Compact Manifolds, Academic Press, 1970.

[W3] C.T.C. Wall, "Classification of hermitian forms II: semi-simple rings", Inv. Math. 18 (1972) 119-141.